W9-ADF-146

Petroleum Refining
in Nontechnical Language

Petroleum
Refining
in Nontechnical Language

t h i r d e d i t i o n
William L. Leffler

Copyright 2000 by
PennWell Corporation
1421 South Sheridan Road
Tulsa, Oklahoma 74112
800-752-9764
sales@pennwell.com
www.pennwell-store.com
www.pennwell.com

Managing Editor: Marla Patterson
Cover design: Brian Firth
Book design & layout: Morgan Paulus

Library of Congress Cataloging-in-Publication Data

Leffler, William L.
  Petroleum refining in nontechnical language / William L. Leffler.-- 3rd ed.
    p. cm
  Rev. ed. of: Petroleum refining for the non-technical person. 2nd ed. c1985.
  Includes index
  ISBN 0-87814-776-4 (hardcover)
      0-87814-867-1 (e-book)
  1. Petroleum--Refining. I. Leffler, William L. Petroleum refining for the
non-technical person. II. Title.

TP690 .L484 2000
665.5'3--dc 21
                                                00-021021

All rights reserved. No part of this book may be reproduced, stored in a
retrieval system, or transcribed in any form or by any means, electronic
or mechanical, including photocopying and recording, without the prior
written permission of the publisher.

Printed in the United States of America
      03 04 05 06   8 7 6 5 4

Contents

List of Figures

List of Tables

List of Acronyms

API	American Petroleum Institute
bbl	barrel
BI	butane isomerization
BS&W	bottom sediment and water
Btu	British thermal unit
BTX	benzene, toluene, and xylene
$C_{11}H_{12}$	alpha-methyl naphthalene
$C_{16}H_{36}$	cetane
C_2H_6	Ethane
C_7H_{16}	heptane
C_7H_{16}	neptane
C_8H_{18}	octane
CCU	catalytic cracking units
CH_3CH_2OH	ethyl alcohol
CH_3OH	methanol
CH_4	methane
CNG	compressed natural gas
CO	carbon monoxide
CO_2	carbon dioxide
cpg	cents per gallon
DEA	diethanolamine
EP	end point
EPA	Environmental Protection Agency
ETBE	ethyl tertiary butyl ether
F	Fahrenheit
FTC	Federal Trade Commission
gal	gallon
H_2	hydrogen
H_2S	hydrogen sulfide
HF	hydrofluoric acid
IBP	initial boiling point

lb/gal	pounds per gallon
lbs	Pounds
LNG	liquefied natural gas
LPG	liquefied petroleum gas
MB/D	Thousand barrels per day
MON	motor octane number
MTBE	methyl tertiary butyl ether
NGL	natural gas liquids
NH$_4$	ammonia
NOx	nitrogen oxides
PONA	Paraffins, olefins, naphthenes, and aromatics
psi	pounds per square inch
psia	pounds per square inch absolute
RON	research octane number
RVP	Reid vapor pressure
S	sulfur
scf	standard cubic feet
SCOT	Shell Claus off gas treating
SMR	steam methane reformer
SO$_2$	sulfur dioxide
SOx	sulfur oxides
SRLGO	straight run light gas oil
TAME	tertiary amyl methyl ether
TBA	tertiary butyl alcohol
TEL	tetraethyl lead
THpEE	tertiary heptyl ethyl ether
THpME	tertiary heptyl methyl ether
THxEE	tertiary hexyl ethyl ether
THxME	tertiary hexyl methyl ether
TNT	tri-nitro-toluene
TOX	toxic compounds
UOP	Universal Oil Products (company)
VOC	volatile organic compound

chapter one
Introduction

The beginning is the most
important part of the book.

-The Republic, Plato

I f you have somehow been charmed by the ads of my publisher or beguiled by the word-of-mouth ravings of a colleague about the efficacy of this book, you probably don't need an introduction to the subject. It's not likely you would have opened this book if you didn't already have at least an uneasy feeling, and maybe even a genuine need to know about petroleum refining. In any event, you have come to the right place.

The layout of the materials in this book is designed to satisfy three different needs. It can be used as a reference book because there's a good table of contents in the front, a good index in the back, and in this edition, a glossary of terms. The book has been used extensively as a text for courses on refining processes. A combination of lectures, reading, and problem solving should have been very reinforcing. Because most people do not have the luxury of listening to a lecturer, the layout is primarily designed for personal study. With that in mind, the dry material has been moistened with as much levity and practicality as could be rendered.

For personal study, the following work plan and comments might be helpful. Chapter 2 on crude oil is most important. Chapter 3 on distilling has a lot of mechanical detail that's not fundamentally important. Don't let it dismay you. The material on vacuum flashing, cat cracking, alkylation, reforming, residue reduction, and hydrocracking are all important as lead-ins to the subject of gasoline. Understanding the mechanics of the refinery gas plants is much less so.

The chapter on gasoline blending can be the most fun (in a nerdish sort of way) because it deals with things familiar yet mysterious—car engines and the stuff that makes them go. For anyone in the administrative part of the business, the chapter on simple and complex refineries will wrap up all the processes into a nice economic package.

The last seven chapters are like *lagniappe*, a small gift a Cajun merchant gives at the time of sale. The information in those chapters is useful but not vital to understanding petroleum refining. So plan to keep your attention span long enough to work through at least the first dozen chapters.

The Evolution of Petroleum Refining

Until the advent of the gasoline engine in the late 19th century, people used petroleum for what we now consider pretty basic needs-heating, light, and maybe lubrication. Even when Colonel Edwin Drake discovered oil at the depth of 69 feet in Titusville, PA in 1859, his investors were excited because they saw an opportunity to compete with whale oil in the illumination market. For the rest of the century, petroleum refining aimed to remove all the light stuff in crude oil that would eventually be used as gasoline. Refiners burned most of it in a pit, just to get rid of it. Initially they were after kerosene for lanterns. They soon recognized the value of the heavier parts of the crude as fuel oil for raising steam and heating buildings—first industrial and eventually commercial and residential.

As a logical consequence, many of the early automobiles, like the famous Stanley brothers' Stanley Steamer, were steam-driven, using kerosene as fuel. By 1890 inventor-entrepreneurs like Karl Benz, Henry Ford, Ransom Olds, and Dave Buick were marketing cars with internal combustion engines that needed a light fuel—gasoline—that changed forever the profile and purpose of petroleum refining.

After 1900 the demand for gasoline exceeded even the volumes formerly burned off in the pits. The chemical engineers then realized they could convert some of the heavier parts of the crude oil by just cooking it until it cracked into lighter products. They invented the first thermal crackers. Even then, the demand for gasoline grew so rapidly that thermal cracking couldn't keep up efficiently. Just as well, since electricity was wiping out the market for lamp oil all through the world and the jet engine hadn't been commercialized yet. Refiners needed a growth product.

Catalysis was still an emerging science, but by 1916 the grandfather of all cat crackers—a fixed bed design—was in place. In 1936 the Frenchman Houdry put the first continuous flow cat cracker on stream. During World War II refiners responded to the demand for high octane (C_8H_{18}) aviation gasoline with alkylation plants and toluene extraction.

Meanwhile the automotive engineers continued to perfect engine design and demanded better quality gasoline. In 1949 the first catalytic reformer went on stream, improving the octane number of the naphtha already being blended into gasoline.

In the latter part of the 20th century, refiners resorted to innovations from the petrochemicals industry to meet both quality and environmental needs. New gasoline blending components made in refineries now had names like methyl tertiary butyl ether (MTBE), tertiary butyl alcohol (TBA), methanol, and ethanol.

Almost all that is changing today is driven by environmental regulation, causing refiners to tweak the existing processes. The technology introduced in the last 15 years has been centered on catalyst improvement, not new processes. All that should be good news to you. It's tough enough to catch up with what's happening until now, without having to worry about what's changing and becoming obsolete after you've learned it.

chapter two

2

Crude Oil Characteristics

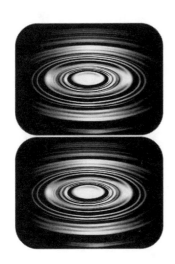

Let these describe the indescribable.

-Childe Harold's Pilgrimage,
Lord Byron

What is crude oil anyway? The best way to describe it is to start by saying what it is not and how it doesn't behave. It is not a single chemical compound—it is a mixture of chemical compounds. The most important of its behavioral characteristics happens as it heats up. When you raise it to its boiling temperature and hold it there, it will not totally evaporate.

Contrast that with water to make a point. Take the pot of water in Figure 2-1 and heat it to 212°F and keep the heat on. What happens? The water starts to boil—it vaporizes or flashes. Eventually, if you keep the heat on, all the water will boil off.

If you had a thermometer in the pot, you would notice that the temperature of the water just before the last bit boiled off would still be 212°F. That's because the chemical compound H_2O boils at 212°F. At atmospheric pressure, it boils at no more, no less.

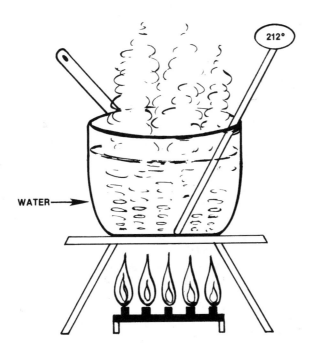

212°

WATER→

Fig. 2-1—Pot of Water

Crude Oil Composition

Now back to crude oil. Unlike water, crude is not one chemical compound but thousands of different compounds. Some are as simple as CH_4 (methane), some are as complex as $C_{35}H_{50}$. CH_4 and $C_{35}H_{50}$ are the chemist's shorthand for individual chemical compounds. No need to get bogged down about that right now. (You can get bogged down with it in chapter 5.) They are all combinations of hydrogen and carbon atoms, called hydrocarbons. The important characteristic, at this moment, is that each of these compounds has its own boiling temperature, and therein lies the most useful and used physical phenomenon in the petroleum industry.

Distillation Curves

To understand a distillation curve, take the same pot and fill it with a typical crude oil. Put the flame to it and heat it up. As the temperature reaches 150°F, the crude oil will start to boil, as in Figure 2-2. Now keep enough flame under the pot to maintain the temperature at 150°F. After a while, the crude stops boiling.

For step two, raise the flame and heat the crude to 450°F. Again the crude starts boiling and after a while stops.

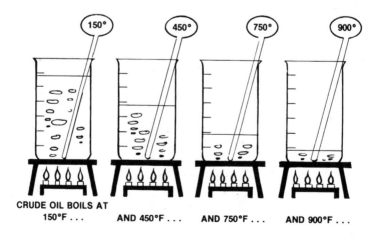

Fig. 2-2—Boiling Temperature of Crude Oil

You could repeat the steps on and on—more and more crude would boil off. What is happening? The compounds that boil below 150°F vaporized in the first step, the compounds that boil at temperatures between 150°F and 450°F vaporized in the second step, and so on.

What you are developing is called a *distillation curve*, a plot of temperature on one scale and the percent evaporated on the other, as in Figure 2-3. Each type of crude oil has a unique distillation curve that helps characterize what kinds of hydrocarbons are in that crude. Generally the more carbon atoms in the compound, the higher the boiling temperature, as shown in the examples below:

Compound	Formula	Boiling Temperature	Weight lbs/gal
Propane	C_3H_8	-44° F	4.2
Butane	C_4H_{10}	31° F	4.9
Decane	$C_{10}H_{22}$	345° F	6.1

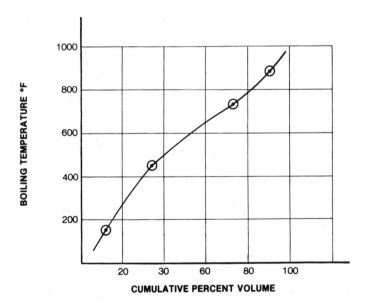

Fig. 2-3—Crude Oil Distillation Curve

Fractions

To further specify the character of crude oil, the refiners have found it useful to lump certain compounds into groups called fractions. Fractions or *cuts* are the generic names for all compounds that boil between two given temperatures, called cut points. The typical crude oil has the following fractions (Fig. 2-4):

Temperatures	Fraction
Less than 90°F	Butanes and lighter
90 - 220°F	Gasoline
220 - 315°F	Naphtha
315 - 450°F	Kerosene
450 - 800°F	Gas Oil
800°F and higher	Residue

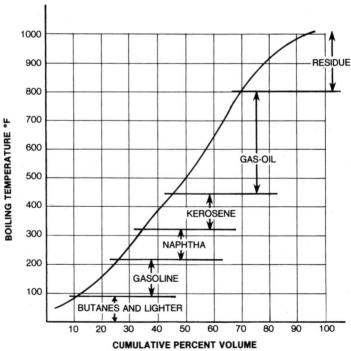

Fig. 2-4—**Crude Oil Distillation Curve and its Fractions**

Later chapters will devote a lot of words discussing the characteristic of each of those fractions, but some are already apparent from their names.

It is important to note that crude oil compositions vary widely. The light crudes tend to have more gasoline, naphtha, and kerosene; the heavy crudes tend to have more gas oil and residue. You might have noticed this from the relationship between the weight of the compounds and the temperature at which they boil. Generally the more carbons, the heavier the compound and the higher the boiling temperature. Conversely, the lower the carbon count, the lighter the fraction and the lower the cut points.

Cutting Crudes

To pull together all this information on distillation curves, follow this quick arithmetic manipulation. Take the curves for the two crudes in Figure 2-5 and run through the steps to determine which crude has higher kerosene content—a bigger kerosene cut.

Kerosene has a boiling temperature range from 315°F to 450°F. Using Figure 2-5, complete the following steps:

1. For the heavy crude—the curve that starts off lower because it has less light hydrocarbons in it—start from the vertical axis at 315°F and intersect the distillation curve, going right to point A. Point A is at 26% on the horizontal axis

2. Now start at 450°F and intersect the same distillation curve, going right, at point B, which is 42% on the horizontal axis

3. Calculate the cumulative volume percentage from the initial boiling point (IBP) of the kerosene to the end point (EP): 42 - 26 = 16%. The heavy crude contains 16% kerosene

4. Now do the same procedure for the light crude and find that 66.5 - 48.5 = 18%

Therefore the light crude has more kerosene in it than the heavy crude. If some refiners were in the business of trying to produce as much

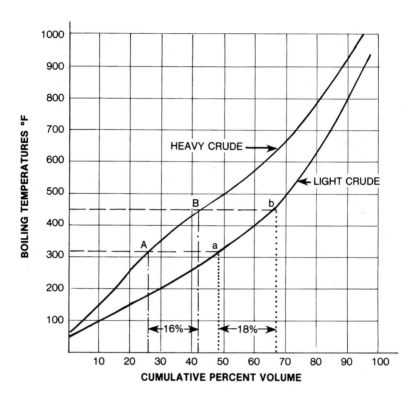

Fig. 2-5—Kerosene Fraction in Two Types of Crude

kerosene as possible, and several occasionally are, they might prefer to buy the lighter crude to the heavier, given a dozen or more other considerations.

Gravities

Gravities measure the weight of a compound, another important characteristic. Chemists always use a measure called specific gravity that relates everything to something universally familiar—water.

The specific gravity of any compound is equal to the weight of some volume of that compound divided by the weight of the same volume of water.

$$Specific\ gravity = \frac{weight\ of\ the\ compound}{weight\ of\ water}$$

The chemists' approach must have been too simple for the petroleum engineers because the popular measure of gravity in the oil industry is a diabolical measure from the American Petroleum Institute (API) called API gravity. This measure is right near the head of the list of measurements with long-forgotten logic. The top of the list is, of course, the Fahrenheit scale for temperature. Really, what was Gabriel Fahrenheit thinking when he set the freezing temperature of water at 32° and boiling temperature at 212°? Why not something easy to remember like 0° and 100° centigrade? For some reason, the formula for API gravity, which is measured in degrees, but has nothing to do with temperature (or angles), is:

$$°API = \frac{141.5}{specific\ gravity} - 131.5$$

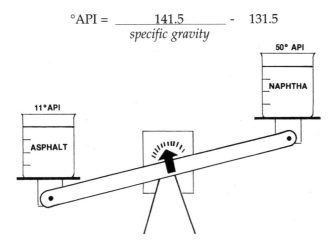

Fig. 2-6—Lower API Gravity, Heavier Liquid

If you play with the formula a little bit, you'll find the following relationships, which might be the mental hooks on which you can hang the concepts:

1. Water has a specific gravity of 1 and an API gravity of 10°
2. The higher the API gravity, the lighter the compound, as shown in Figure 2-6

3. The reverse is true for specific gravity

Typical gravities:

	API Gravity
Heavy Crude	18°
Light Crude	36°
Gasoline	60°
Asphalt	11°

Common knowledge says oil floats on water. The sheen you might have seen from a boat or dock results from oil not dissolving in water and being at an API gravity above 10°. Not all oil weighs that little. Industry lore has stories of barge operators who assumed that all oils are lighter than water. To their horror as they filled their barge with heavy fuel oil, it sank before their eyes. After the fact, they learned they were loading 9° API fuel oil.

Sulfur Content

One more excursion on the subject of crude oil is appropriate at this point—a discussion of the sulfur content in crude oil. One annoying aspect nature endowed on crude oils is the differing amounts of sulfur content in various types of crude oil. To complicate this bequest, the sulfur is not in the form of elemental sulfur—a chemical all by itself—but is usually a sulfur compound. It is chemically bonded to some of the more complicated hydrocarbon molecules so that it is not easily separated from the pure carbon compounds, i.e., not until it is burned. It will then form one of several smelly or otherwise environmentally objectionable sulfur/oxygen or sulfur/something compounds. So sulfur removal before hydrocarbons get to the burner tip remains a big issue for refiners today and will be the subject of more words later on.

Americans generally spell sulfur with an f, while the British use a ph. Both are acceptable, but as usual the Yanks are more efficient.

The parlance in discussing crude oils of varying sulfur content is to categorize them into sweet crudes and sour crudes. This quaint, faintly oriental designation of sweet and sour has more to do with taste than you might think. In the early days of Pennsylvania crude oil production, petroleum was primarily sought to make kerosene as a substitute for the whale oil used as lamp oil for indoor lighting. If a kerosene fraction had too much sulfur, it would have an unacceptable smell when it burned. In addition, the sulfur would accelerate the rate at which silver would tarnish—clearly a bad thing to have in the house. Somewhere along the line, someone discovered that kerosene with higher sulfur content had a more sour taste and that with a low sulfur content had a sweeter taste. Over a time long enough for the designation to become permanent, tasting was generally the acceptable method for determining which crudes would make good lamp oil.

Today, sweet crudes typically have 0.5% sulfur content or less, sour 1.5% or more. The area in between is sometimes called intermediate sweet or intermediate sour, but the distinction is not clear. What may be sweet to some may be sour to others, now that refiners have no more tasters around.

Typical sweet crudes are West Texas Intermediate (the popular, traded crude on the New York Mercantile Exchange), most Louisiana and Oklahoma crudes, North Sea, and Nigerian crudes.

Sour crudes include Alaska North Slope, Venezuelan, and West Texas Sour from fields like Yates and Wasson.

Intermediate crudes include California Heavy, such as from the San Joaquin Valley and many of the Middle East crudes.

Volumes

One other convention grew out of the Pennsylvania oil fields, according to oil patch lore. Oil was initially shipped to market by horse-drawn wagon or flatcar in used 46, 48, and 50 gallon (gal) wine barrels. Refiners insisted on allowing for spillage and leakage after producers hauled it across not-so-great roads from the oil fields to the refineries. They paid on the basis of 42 gals, which became the "oil barrel."

Transportation in the US was primarily by wagon, train, and eventually truck and pipeline. Measuring the volume made sense. In the rest of the world, particularly in Europe, the industry moved oil by seagoing vessels. That required calculating the weight of the cargo to assure the load would not exceed the displacement of the ship. King Henry VIII found out the mechanics of displacement when he overloaded his flagship, the *Mary Rose*, with cannon and troops, and watched it sink off Portsmouth Harbor as it set off to battle the French in 1545. In Europe and Asia, weight usually measured in tonnes (2240 US gals) became the maritime standard by which oil was bought, sold, and transported. Curiously enough, gasoline at the retail pump has always been sold everywhere by volume, in gallons or liters, more because of the meters than anything else.

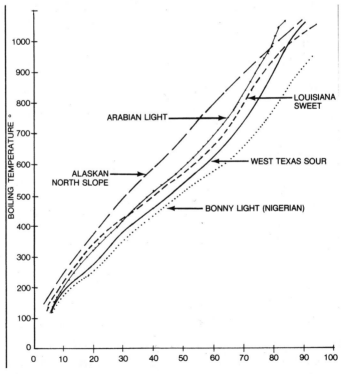

Fig. 2-7—Distillation Curves for Some Crude Oils

Review

Figure 2-7 shows the distillation curves for five different crudes: three US and two non-US. Some have more light fractions, some more heavy. All have different prices, so refiners will have different incentives to process them, particularly when they want different cuts.

Exercises

1. Draw the distillation curve for the following crude oils (on the same graph):

	% volume	
Lighter than	Oklahoma Sweet	California Heavy
113°F	5.1	-
113-220	9.2	-
220-260	4.0	-
260-315	5.7	4.2
315-390	9.3	5.1
390-450	5.4	4.8
450-500	5.8	8.5
500-550	4.7	7.9
550-650	10.8	8.0
650-750	8.6	14.8
750-900	13.5	15.1
900-1000	5.9	13.4
more than 1000	12.0	18.1

How much naphtha (220-315°F) is there in each crude?

2. Suppose you had a beaker of 11°API asphalt and a beaker of 50°API naphtha, both equal volumes. If you mixed them together, what would be the resulting API gravity? The answer is not 30.5° API.

c h a p t e r   t h r e e

Distilling

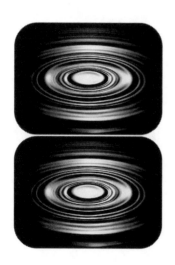

Why should we rise, because 'tis light?

-Break of Day, **John Donne**

A casual passerby of a refinery can make an easy mistake by referring to the many tall columns inside as "cracking towers." In
fact, most of them are distilling columns of one sort or another.
Cracking towers, which are usually shorter and squatter, will be covered
in a later chapter.

Distilling units are the clever invention of process engineers who
exploit the important characteristic discussed in the last chapter—the distillation curve. The mechanism they use is not too complicated but, for
that matter, not all that interesting. However, in the interest of complete-
ness and familiarity, you can cover the rudiments here.

The Simple Still

For years Kentucky moonshiners used the simple still in Figure 3-1 to
separate the white lightning from the dregs. After the sour mash fermented—i.e., a portion of it had slowly undergone a chemical change to
alcohol—they heated it to the boiling range of the alcohol. The white
lightning vaporized—as a vapor, it is less dense (lighter) than liquid. It
moved out of the liquid, then through the condenser where it cooled and
turned back to liquid. The liquid left in the still was discarded. The liquid
that ended up in the condenser was bottled. A process engineer would
call this a simple *batch process distillation.*

If the moonshiners wanted to sell a better-than-average product, they
might have run the product through a second *batch still* much like the
first. There they could have separated the best part of the liquor from
some of the non-alcoholic impurities that inevitably flowed along with
the overhead in the first still. Some impurities might have gone overhead
because of the inadequacy of the temperature controls or because the
moonshiners wanted to be sure they got all they could so they set the
temperature a little high on the first batch.

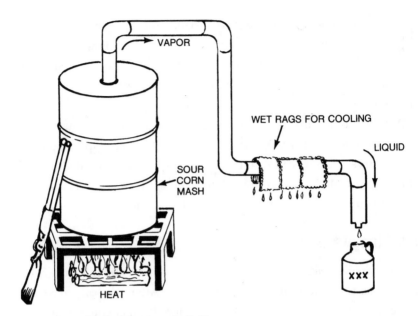

Fig. 3-1—**The Moonshiner's Still**

This two-step operation could be made into the continuous operation shown in Figure 3-2. In fact, many early oil distilling operations looked like that.

The Distilling Column

The two-step batch distilling operation is obviously not suited for handling several hundred thousand barrels per day of crude oil with five or six different components being separated. A *distilling column* can do it on a continuous basis with fewer facilities and much less labor and energy consumption.

Figure 3-3 shows from afar what happens at a crude distilling column. Crude goes in, and the products go out—gases (butane and lighter), gasoline, naphtha, kerosene, light gas oil, heavy gas oil, and residue.

What goes on inside the distilling column is more complicated. The first piece of equipment important to the operation, the *charge pump*,

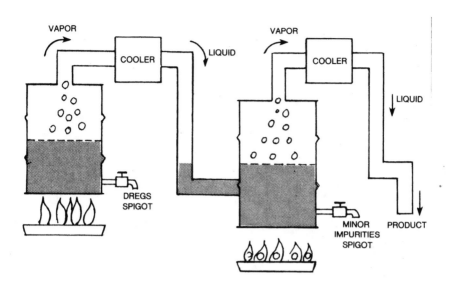

Fig. 3-2—**Two Stage Batch Still**

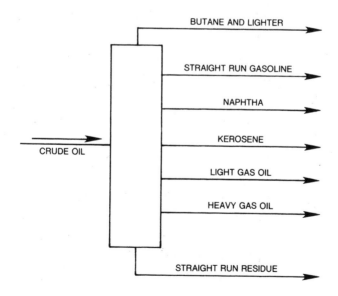

Fig. 3-3—**Distilling**

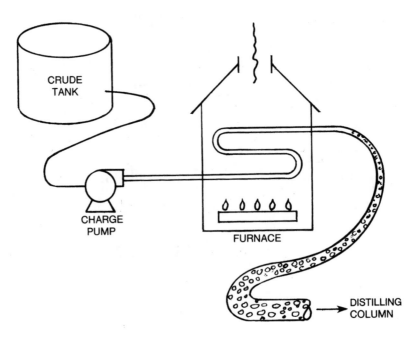

Fig. 3-4—Crude Oil Feed to Distilling

moves the crude from the storage tank through the system (Fig. 3-4). The crude is first pumped through a furnace where it is heated to a temperature of about 750°F. From the knowledge developed in the last chapter, you can see that more than half of the crude oil changes to vapor form as the furnace heats it to this temperature. This combination of liquid and vapor is then introduced to the distilling column.

Inside the distilling column there is a set of *trays* with perforations in them. The perforations permit the vapors to rise through the column. When the crude liquid/vapor charge hits the inside of the distilling column, gravity causes the denser (heavier) liquid to drop toward the column bottom, but the less dense (lighter) vapors start moving through the trays toward the top, as Figure 3-5 shows.

The perforations in the trays are fitted with a device called *bubble caps* (Fig. 3-6). Their purpose is to force the vapor coming up through the trays

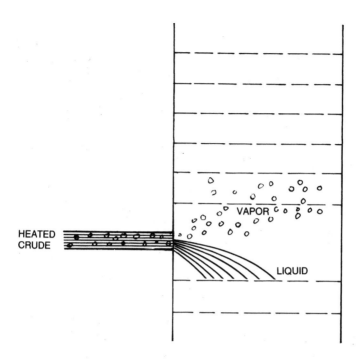

HEATED CRUDE

VAPOR

LIQUID

Fig. 3-5 — Crude Entering the Distilling Column

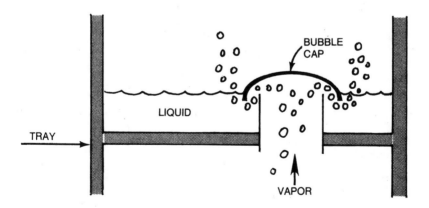

BUBBLE CAP

LIQUID

TRAY

VAPOR

Fig. 3-6—Bubble Cap on Distilling Column

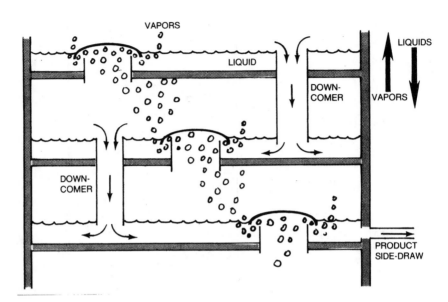

Fig. 3-7—**Downcomers and Sidedraws**

to bubble through the liquid standing several inches deep on that tray. This bubbling is the essence of the distilling operation: the hot vapor (starting at 750°F) bubbles through the liquid. Heat transfers from the vapor to the liquid during the bubbling. As the vapor bubbles cool a little, some of the hydrocarbons in them will change from the vapor to the liquid state; the temperature of the vapor drops and the lower temperature of the liquid causes whatever heavier compounds remain in the vapor to condense (liquefy).

After passing through the liquid and shedding some of the heavier hydrocarbons, the vapor then moves to the next tray where the same process takes place.

Meanwhile, the amount of liquid on each tray is growing as some of the hydrocarbons from the vapor are stripped. Figure 3-7 shows a device called a *downcomer*, installed to permit excess liquid to overflow to the next lower tray. At several levels on the column, the *sidedraws* shown take the liquid distilled product off—the lighter products from the top of the column, the heavier liquids toward the bottom.

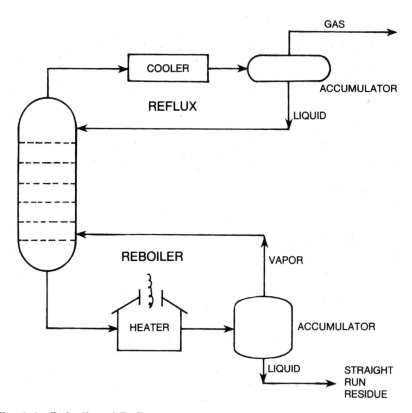

Fig. 3-8—Reboil and Reflux

Some molecules actually make several round trips: up several trays as vapor, finally condensing, then down a few trays via the downcomer as a liquid. It's this vapor-liquid mutual scrubbing that separates the cuts—once through won't do it.

Reflux and Reboil

Several things go on outside the distilling column that facilitates the operation. To assure some of the heavies don't get out the top of the column, occasionally some of the vapor will be run through a cooler. Whatever is condensed is reintroduced to a lower tray. Remaining vapor is sent off as product. The process is a form of *refluxing* (Fig. 3-8).

Conversely, some of the light hydrocarbon could be entrained on the bottom of the column where the liquid part of the crude oil ended. So a sidedraw may be used to recirculate the liquid through a heater to drive off any lighter hydrocarbons for reintroduction as a vapor at some lower level in the distilling column. This is called *reboiling*. The advantage of a reboiler is that only a small fraction of the crude stream has to be worked on to accomplish the additional recovery. The whole crude oil stream volume doesn't have to be heated and pumped, so the reboiler saves energy and money.

Reboiling or refluxing can be used effectively in the middle of the column as well, facilitating good separation. The reboiler has the advantage of giving heat input to help push lighter molecules up the column. Similarly, refluxing can give a last shot at condensing some heavy molecules that may have gotten too high in the column.

Sometimes the crude composition is such that there would not be enough liquid/vapor flow on some of the trays of the column. A very light crude might not have anything coming out the bottom or the lower side draws. More often refiners run into the converse when they buy very heavy crudes like Mayan crude from Mexico or some Venezuelan crudes They have very little, if any, light hydrocarbon to draw off the top of the column. In that case, using reflux and reboil can regulate the flows to keep the distilling (separating) operation going. As an alternative, refiners sometimes blend the "extreme" crudes with other crudes, if the reboil or reflux capacity is insufficient. That seems ironic since they are mixing things together so they can separate them better, but that's one way to extend the capability of existing hardware.

Cut Points

For analyzing and controlling distilling operations, the key parameters are *cut points*, the temperatures at which the various distilling products are separated. The temperature at which a product (or *cut* or *fraction*) begins to boil is called the *initial boiling point* (IBP).

The temperature at which it is 100% vaporized is the *end point* (EP). So every cut has two cut points, the IBP and the EP. The diagram back in Figure 3-3 makes it readily apparent that the EP of naphtha is the IBP of

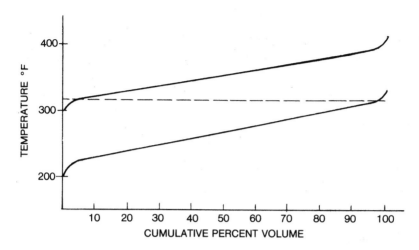

Fig. 3-9—Overlapping Distillation Curves

kerosene. At a cut point, the EP and IBP of the two adjacent cuts are nom-inally the same. That depends on how good a separation job the distilling column does. The IBP of the kerosene could be higher than the EP of the naphtha. You may have wondered, looking at the mechanics of trays and bubble caps, how well the process works. In fact, the operation is a little sloppy, resulting in what is referred to as, pardon the expression, *tail ends*.

For instance, in a laboratory analysis of the naphtha and kerosene, the two distillation curves would look like the curves in Figure 3-9. Looking closely, the EP of naphtha is about 325; the IBP of the kerosene is about 305.

A better way of visualizing the tail ends is in Figure 3-10, which plots temperature—this time not against cumulative percent vaporized but against percent being vaporized at that temperature (the first derivative of the above diagram, if you know calculus).

Almost all refining operations generate the phenomenon of tail ends. It is so common, everyone takes it for granted. However, to simplify analysis, most people make an accommodation. Refiners usually mean *effective* cut points when they talk about distillation. These represent the compromise temperature at which the cuts can be considered cleanly cut. Typically 1% might evaporate lower than the IBP and higher than the EP.

In the petrochemicals business, producers deal with single chemicals and delicate chemical reactions. Variations like this cannot be tolerated, but most end uses of petroleum are more accommodating. When you see *cut point* in the rest of this book, it will implicitly mean *effective cut point*.

Setting Cut Points

You might have taken the cut points referred to in the last chapter and in the discussions so far as an immutable given for naphtha, kerosene, etc., (even considering the little discussion of the difference between cut points and effective cut points). There is some latitude in setting the cut points on a distilling column. Changing the cut point between naphtha and kerosene has several implications. If the cut point were changed from 315 to 325, several things would happen. First, the *volumes* coming out of the distilling column would change—more naphtha, less kerosene. That's because the fraction between 315 and 325 would then come out the naphtha spigot instead of the kerosene spigot.

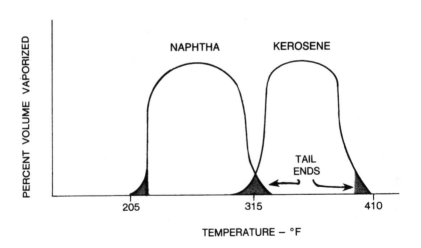

Fig. 3-10—**Tail Ends in Distillation Curves**

At the same time the *gravities* of both naphtha and kerosene get heavier. How can that be? The cut moving out of kerosene into naphtha is heavier than the average naphtha gravity. It is also lighter than the average kerosene gravity. So both cuts get heavier!

That might remind you of the comment by the cynical coach who just lost a player to a rival. "I figure the quality of both teams just improved."

Some of the other properties also change, but gravity is the only one that has been discussed so far. When the operations downstream of the distilling unit are discussed in later chapters, you'll appreciate the further implications of changing cut points on the distilling unit.

The listing shown in Figure 3-11 of all the distilling unit cut destinations will help set up future chapters. The light ends from the top of the column (the overhead) go to the gas plant for separation. The straight run gasoline goes to motor gasoline blending or perhaps isomerization. Naphtha goes to the cat reformer for upgrading to a better quality gaso-

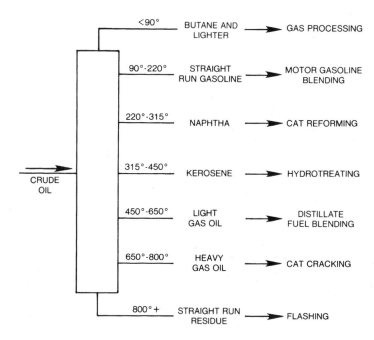

Fig. 3-11—Distilling Crude and Product Disposition

line blending stock. Kerosene goes to a hydrotreater for cleanup. Light gas oil goes to distillate fuel oil blending. Heavy gas oil goes to the cat cracker as feed. And straight run residue is used for heavy fuel oil or is fed to the flasher.

Exercises

1. Fill in the blanks below with nomenclature from the following list:

tail ends	downcomer	overhead
IBP	EP	cooler
furnace	straight run gasoline	continuous
crude oil	fractionator	
batch	bubble cap	
increases	decreases	

   a. After moonshine comes out the top of a still, it has to be run through a _____ before it can be bottled.

   b. _____ processing is not very practical in a modern refinery. Today all crude distillation is a _____ operation.

   c. The holes in the distilling column trays are fitted with either a _____ or a _____.

   d. The device that enables the essential job of scrubbing in a distilling unit is the_____ _____.

   e. Tail ends occur because the _____of one cut overlaps the _____ of another.

   f. As vapor moves up the distilling column, its temperature _____.

   g. Lowering the EP of a distilling column cut _____ the volume of that cut and _____the API gravity.

2. The refinery operations manager is told he must produce 33 thousand barrels per day (MB/D) of furnace oil this winter. He is also told he will be supplied with 200 MB/D of crude—30 MB/D of Louisiana crude

and 170 MB/D of West Texas. The distillation curves for those crudes are shown below. Another premise in the operation is that he must maximize jet fuel., i.e., he must squeeze as much out of the crude as possible. He knows that the boiling range for jet fuel is 300-525°F, so he will have to set the cut points on his distilling units at these two temperatures.

Finally in order to meet the specifications for the 33 MB/D of furnace oil, 20 MB/D of straight run light gas oil (SRLGO) will have to be cut from the crude on the distilling units and provided to the furnace oil blending operation.

Problem: at what temperatures will the operations manager have to cut SRLGO to make 20 MB/D?

Distillation Data:

	West Texas % Volume	Louisiana % Volume
IBP-113°F	15	20
113-260	12	18
260-315	18	15
315-500	10	15
500-750	20	12
750-1000	10	10
more than 1000	15	10

Hints:

- Calculate a composite crude oil distillation curve
- The turbine fuel EP is the SRLGO IBP
- Calculate the SRLGO EP that gives 20 MB/D

Save the answer to this problem. Problems in subsequent chapters will build on this exercise.

chapter four
4
Vacuum
Flashing

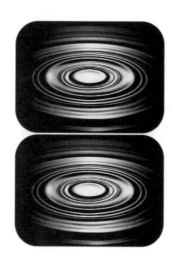

A vacuum is repugnant to reason

- Principles of Philosophy
Rene Descartes

T he discussions so far about distillation curves and distilling columns have treated the shape of the curves at temperatures close to 900°F somewhat vaguely, but on purpose. There is a phenomenon that happens at these temperatures called cracking. Refiners use cracking extensively, but in controlled ways. In the distillation process, all the uncontrolled symptoms of cracking could occur. To avoid them, refiners developed vacuum flashing.

The Cracking Phenomenon

Suppose a technician were going to develop a crude distillation curve in the laboratory. Technicians do that all the time to prepare *assays* for use by refiners to help assess the suitability of crudes. The technician would heat the crude oil, note and record the temperature, capture the vapor and condense (liquefy) it, and record and cumulate the volume of the liquids. If the technician permitted the temperature of the crude to rise to about 900°F, a sudden change in the shape of the distillation curve would take place. As the temperature went from 900 to 1000°F, the cumulative volume recovered would exceed 100%—and crude would still be boiling in the pot!

This might exaggerate what anyone can easily observe with the naked eye, but it makes the point. As the complicated hydrocarbon molecules that have not yet vaporized at 900°F are heated to higher temperatures, the energy transfer from the heat is enough to cause the molecules to crack into two or more smaller molecules. For example, in Figure 4-1 a molecule of $C_{16}H_{34}$ may crack into three pieces, C_8H_{18}, C_6H_{12}, and C_2H_4. (A fuller discussion of the chemical reaction of cracking will be covered in the chapter on cat cracking).

If you recall the discussion on boiling points, it starts to become apparent why the shape of the distillation curve behaves so strangely at this point. The smaller molecules boil at much lower temperature than the larger ones. As cracking creates them, they leap out of the crude oil. That tells only half the story. Why does more volume get created so that the distillation curve exceeds 100%? In simplest terms, the smaller molecules take up more room than the larger ones.

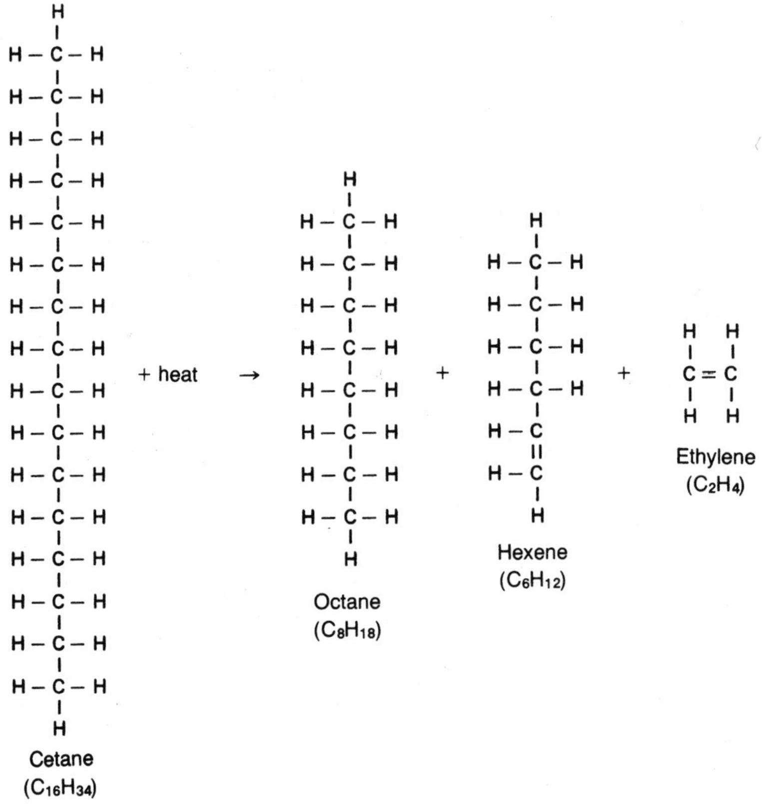

***Fig. 4-1*—Molecule Cracking**

Again, recall the discussion on hydrocarbon properties and specific gravities. The compound $C_{16}H_{34}$ weighs 7.2 pounds per gallon (lb/gal) and C_8H_{18}, C_6H_{12}, and C_2H_4 weigh 5.9, 5.6, and 3.1, respectively. Take a gallon of $C_{16}H_{34}$ and suppose you could crack it completely into these three compounds. (It never happens that way—combinations of different compounds turn up too). The chemistry of it says that there would be 50% C_8H_{18}, 38% C_6H_{12}, and 12% C_2H_4. The percentages are by weight, not by volume (Fig. 4-2). To make the calculation easy, start with 7.2 pounds (lbs) of $C_{16}H_{34}$ and calculate the change as follows:

cracking 100% *of* $C_{16}H_{34}$ = 7.2 lb ÷ 7.2 lb/gal = 1.00 gal

results in

50% yield of C_8H_{18} = 3.6 lb ÷ 5.9 lb/gal = 0.61 gal
38% yield of C_6H_{18} = 2.7 lb ÷ 5.6 lb/gal = 0.48 gal
12% yield of C_2H_4 = 0.9 lb ÷ 3.1 lb/gal = 0.29 gal
————————————————————————————————————
100%   7.2 lb   1.38 gal

So, 7.2 lbs of C_2H_4 take up 1 gal, but 7.2 lbs of the components take up 1.38 gals. This phenomenon exists because the larger molecules tend to have the atoms packed closer together than the smaller molecules. The larger ones take up less space per lb.

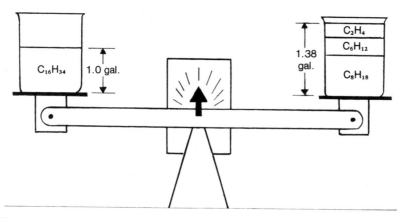

Fig. 4-2—$C_{14}H_{34}$ and its Cracked Components

Cracking is a lucrative process only when it is controlled. The distilling unit is not designed to control it, so refiners avoid it by not raising the temperatures in the distilling process. The heaviest cut point on the distilling column is around 750°F. There are, nonetheless, lots of hydrocarbons that need to be separated from the straight run residue and the technique used is vacuum flashing.

Effects of Low Pressure

Suppose you own a chain of restaurants that specialize in serving stuffed baked potatoes—one in Houston, Texas and one in Denver,

Colorado. Do you know that baking potatoes in Denver takes 20 minutes longer than in Houston? (I found that out the hard way when I went skiing in the Rocky Mountains and cooked my own supper one night).

The reason is that at the higher altitude of Denver, the atmospheric pressure is lower. When people say the air in the mountains is thinner, they really mean there is less air per cubic inch. Since the pressure is lower, water will boil at a lower temperature. Even though you have the oven set at 375°F, the boiling moisture will keep the potato at the lower temperature of 205°F in Denver, not the 212°F in Houston. Thus the potato doesn't ever get as hot in Denver as it does in Houston and it takes longer to cook.

All this illustrates the simple relationship between pressure and temperature. The process of heating lets the molecules absorb enough energy to escape from the liquid form to the gaseous form. The rate at which they escape depends on the rate at which the heat is delivered (how high you turn the burner) and the pressure of the air above them. The lower the pressure, the less energy has to be transmitted, and the lower the temperature at which the vapor will start forming in the liquid, *i.e.*, boil.

The point is, the lower the pressure, the lower the boiling temperature.

Vacuum Flashing

Apply the pressure/boiling relationship to the crude oil cracking problem. The straight run residue will crack if the temperature goes too high, but straight run residue needs to be separated into more cuts. The solution is to do the distillation at reduced pressure.

In the vacuum flasher shown in Figure 4-3, straight run residue from the distilling unit, while it is still hot, is pumped to the flasher. The flasher is a large diameter vessel, too squat to be called a column, but that's more or less what it is. The straight run residue leaves the pump and heads down the line and into the flasher, where the pressure has been lowered well below atmospheric pressure. At that low pressure the lighter portions of the straight run residue will vaporize, but at that temperature, no cracking will take place.

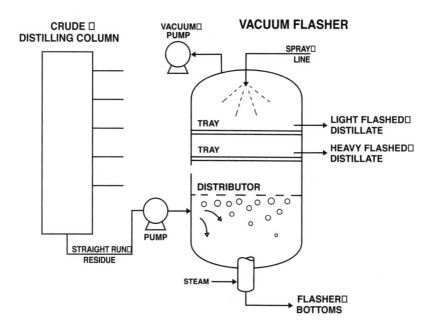

Fig. 4-3—**Vacuum Flashing**

Various gadgets inside the flasher facilitate separating and capturing the offtake streams. As the mixed stream of vapor and liquid enter the flasher, almost all the liquid falls to the bottom. Some of the liquid forms droplets—not quite gaseous, but carried along with the vaporized part of the straight run residue. They start to rise with the vapor. To capture these droplets and to disperse the vapor evenly throughout the flasher, the vapor/droplet mixture encounters a *distributor*—a thick mesh screen or a tray of loose metal rings several inches thick. The distributor catches the droplets, which drip down to the bottom. Further up the flasher are two trays that operate the same as those in a distilling column. There will probably also be a reboiler, though it's not shown in the figure. Reflux, serving the same purpose as it does in a distilling column, is in the form of a liquid spray from the top of the vessel, using some of the cooled streams drawn off one of the trays.

The *vacuum pump* at the top of the vessel maintains the low pressure in the vessel and continuously draws off any vapors that have not condensed, and usually consist of small amounts of water and some hydrocarbon. Some vacuum flashers use a device called a steam eductor to maintain the low pressure.

The two streams drawn off the upper part of the flasher, *light flashed distillate* and *heavy flashed distillate*, together are called *flasher tops*. The two streams are often kept segregated. Refiners found through chemical analysis that the heavier part of the flasher tops contains most of the contaminants that poison catalysts in processing units further downstream. By keeping the streams segregated, they can treat the heavy flashed distillate and remove most of the bad actors.

Meanwhile, at the bottom of the flasher, the liquids drain through a pipe. Some flashers have apparatus for one last pass at removing the good stuff. They pump some super-heated steam up through the pipe as the liquid comes down, releasing some lighter hydrocarbons that may have been entrained in the liquid. Because the superheated steam is at temperatures well above the cracking temperatures, the liquid coming out of the bottom, the *flasher bottoms*, is quenched with some cooled liquids before the cracking has a chance to take place. (Later, in the chapters on cat cracking and coking, you'll find out that cracking requires a minimum *residence time* to get started).

The flasher bottoms have several destinations—feed to an asphalt plant, a thermal cracker, a coker, or as a blending component for residual fuel. The real reason to run a flasher is to make the flasher tops to feed to a cat cracker.

Review

Flashing is the equivalent of distilling the straight run residue at a cut point in the neighborhood of 1000-1100°F. Crude oil distillation curves always show the temperature-volume relationship as if the theoretical distillation took place. *i.e.*, they assume the vacuum flasher is part of the distillation unit.

Exercises

1. Fill in the blanks:

    a. Vacuum flashing is a technique to continue _____ of crude oil at high temperatures without encountering _____.

    b. The lower the pressure around a liquid, the _____ (lower, higher) the boiling temperature.

    c. The higher the pressure in a vacuum flasher, the _____ (lower, higher) the EP of the flasher tops, assuming the feed rate and feed temperature don't change.

    d. Products from a flasher are _____, _____, and _____.

    e. To reduce the volume of flasher tops, holding the feed rate constant, you either _____ the pressure in the flasher or _____ the temperature of the flasher feed.

2. Use the data given to the operations manager in the previous chapter. In order to satisfy the need for feed to the asphalt plant, 35 MB/D of flasher bottoms must be made available.

    a. At what maximum cut point does the operations manager set the flasher to assure supply to the asphalt plant? Use the distillation curve temperatures, not the internal flasher temperatures, since you don't know at what pressure the flasher operates.

    b. If the straight run residue was cut from the distilling column at 800°F, what would be the volume of the flasher tops?

chapter five

The Chemistry of Petroleum

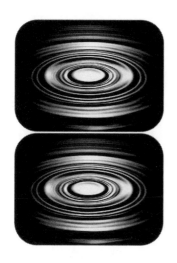

It strikes me that Nature's system must be a real beauty, because in chemistry we find that the associations are always beautiful whole numbers—there are no fractions.

-R. Buckminster Fuller

S o far, in most of the discussion of crude oil and refining you have been able to avoid the basics of chemistry. The good times have come to an end. Distilling and flashing are basically mechanical processes, but the rest of the refining processes, for the most part, are chemical reactions. While you don't have to have taken Chemistry I and have gotten at least a C in Physical Chemistry, you will have to know something about hydrocarbons to go on from here.

Don't skip this chapter! You will have to keep a figurative thumb in these pages if you want a full appreciation of the subsequent seven chapters.

Atoms and Molecules

Physicists are continually surprising the scientific world by finding smaller and smaller particles, even more esoteric than neutrons and electrons. Luckily, the smallest, useful notion in petroleum refining is the *atom*. Examples are carbon, hydrogen, sulfur, and oxygen, whose chemical symbols are C, H, S and O.

The characteristics of matter depend on the types of atoms involved and how they are attached to each other in groups called *molecules*. There are rules by which atoms can be arranged into molecules. The most important rules have to do with *valences* and *bonds*.

- *Valences.* Each type of atom (element) has an affinity for other elements according to its atomic structure. For example, carbon atoms always like to attach themselves to four other atoms. Hydrogen atoms always like to attach themselves to only one other atom.

  Definition: The valence of an atom of any element is equal to the number of hydrogen atoms (or their equivalent) with which the atom can combine.

- *Bonds.* The connection between two atoms is called a bond. You can think of it as an electrical force that ties the two atoms together.

Hydrocarbons

The simplest example of valence, bonds, and hydrocarbons is methane, which has one carbon and four hydrogen atoms. Look at the structure in Figure 5-1, and see all the valence rules are satisfied.

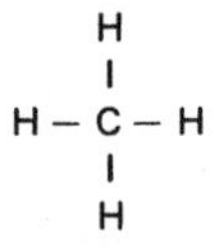

$$
\begin{array}{c}
\text{H} \\
| \\
\text{H} - \text{C} - \text{H} \\
| \\
\text{H}
\end{array}
$$

Methane
(CH_4)

Fig. 5-1—Methane

If you take the next simplest hydrocarbon, ethane, some further complications can be introduced. Ethane is C_2H_6. You can check that each carbon has four attachees; each hydrogen has one (Fig. 5-2). Note that one carbon atom is attached to another one, each one using the same bond to satisfy its valence need. That's okay. Similarly, whenever you see hydrogen referred to as a separate compound, it will always be called hydrogen because that is how it exists—one hydrogen atom attached to a second, satisfying the valence rule of both atoms.

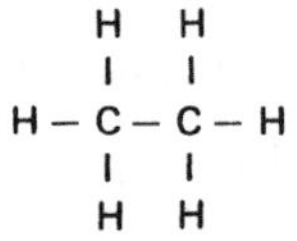

$$
\begin{array}{c}
\text{H} \quad \text{H} \\
| \quad\ | \\
\text{H} - \text{C} - \text{C} - \text{H} \\
| \quad\ | \\
\text{H} \quad \text{H}
\end{array}
$$

Ethane
(C_2H_6)

Fig. 5-2—Ethane

A whole class of hydrocarbons can be defined by extending the relationship from methane to ethane and beyond. These molecules are *paraffins* or *straight chain compounds* and have the formula C_nH_{2n+2}. The examples in Figure 5-3 are propane, normal butane and normal pentane.

```
        H   H   H
        |   |   |
  H  -  C - C - C  - H
        |   |   |
        H   H   H
```
Propane
(C₃H₈)

```
        H   H   H   H
        |   |   |   |
  H  -  C - C - C - C  - H
        |   |   |   |
        H   H   H   H
```
Normal Butane
(C₄H₁₀)

```
        H   H   H   H   H
        |   |   |   |   |
  H  -  C - C - C - C - C  - H
        |   |   |   |   |
        H   H   H   H   H
```
Normal Pentane
(C₅H₁₂)

Fig. 5-3—Propane, Normal Butane, Normal Pentane

Why the prefix "normal" before butane and pentane? Well, there are several ways to arrange the carbon atoms in C_4H_{10} and C_5H_{12}. One of them is shown in Figure 5-3. Another is to put a branch off one of the inside carbons. In that case, the compounds C_4H_{10} and C_5H_{12} are called isobutane and isopentane (Fig. 5-4).

Even though normal butane and isobutane have the same formula, they behave differently. They boil at different temperatures; they have different gravities (since they are packed differently); and they cause different chemical reactions, which will be important in a later chapter on alkylation.

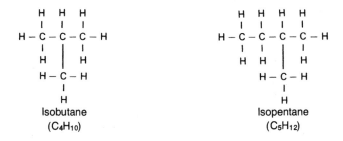

Isobutane
(C₄H₁₀)

Isopentane
(C₅H₁₂)

Fig. 5-4—Isobutane, Isopentane

There is a shorthand convention used for these light hydrocarbons. When you talk about mixtures of streams that have, for example, only propane, ethane, and maybe methane but no butanes nor larger (heavier) molecules, you can call that mixture or stream propane and lighter, or C_3 minus, or C_3^-. Also, you can say it contains no butane and heavier or C_4^+. The convention is used for any of the light hydrocarbons, mostly through C_5 or C_6.

Naphthenes

Another class of hydrocarbons found in petroleum has either five or six carbons in them. The carbons are bent into a ring or cyclic shape, as in Figure 5-5, giving rise to the name cyclopentane (C_5H_{10}) and cyclohexane (C_6H_{12}). Note that there is less hydrogen in cyclopentane than in normal or isopentane and the same for the C_6's. This class of compounds is called *naphthenes*.

Beyond the simple paraffins and cyclic hydrocarbons an infinite variety of possibilities occur by connecting two or more types of compounds. The least complicated example is methyl cyclopentane, which is the connection of $-CH_3$, called a *methyl radical*, and a cyclopentane at the spot

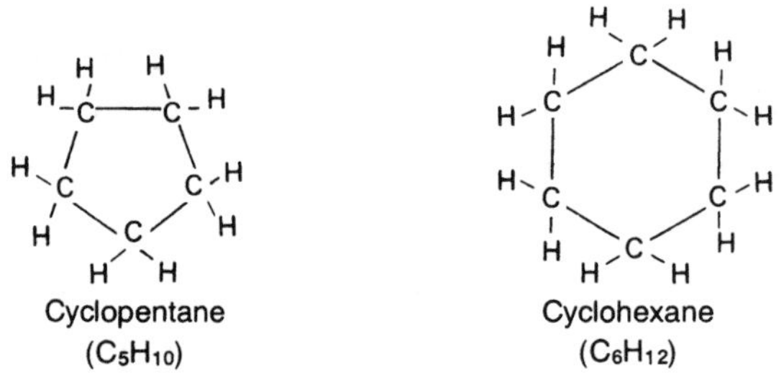

Cyclopentane
(C_5H_{10})

Cyclohexane
(C_6H_{12})

Fig. 5-5—Naphthenes

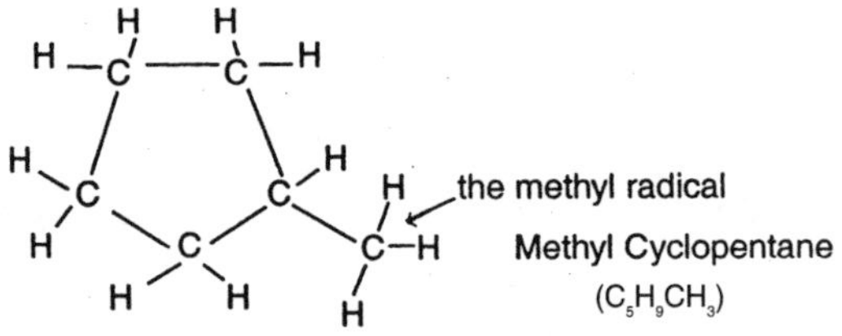

the methyl radical
Methyl Cyclopentane
($C_5H_9CH_3$)

Fig. 5-6—Methyl Cyclopentane

where hydrogen ought to be, forming C_6H_{12} as in Figure 5-6. Actually, a more explicit chemical formula for methyl cyclopentane is $C_5H_9CH_3$, which better shows the methyl and the cyclopentane radicals.

> *A radical is a molecule wannabe. An ethyl radical, for instance, would be $-C_2H_5$, and wants to be a C_2H_6 molecule but is missing one hydrogen. Radicals are transient forms of compounds that are highly reactive because their valences are unsatisfied. During chemical reactions, they abound momentarily until they find partners and join up, forming new compounds. Refiners and chemists find it useful to refer to radicals as if they were stand-alones to explain the intermediate steps of a process.*

Olefins and Aromatics

It is possible to have a compound with two carbons and only four hydrogens. That would seem to violate the valence rules you just spent six minutes learning. But in the chemical compound ethylene in Figure 5-7, there is a *double bond* between the two carbon atoms to make up for the deficiency of hydrogen atoms. Ironically, the double bond holding the two carbons together is weaker, not stronger, than a single bond. You can think of it as two bonds squeezed into the space of one. The compound is therefore chemically unstable and can be reacted with some other compound or element with relative ease, eliminating the double bond. That is why ethylene is such a popular compound for making other compounds.

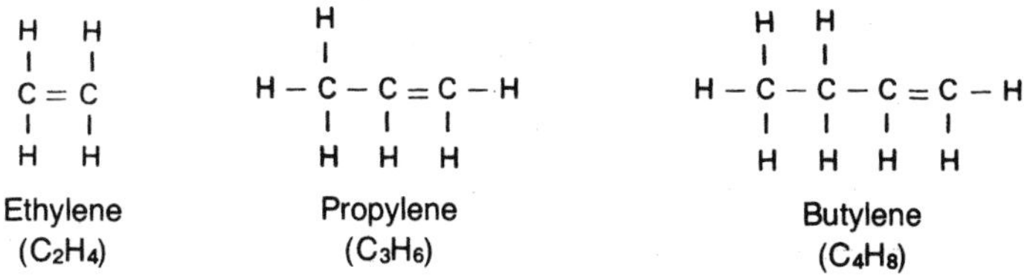

Ethylene (C_2H_4)   Propylene (C_3H_6)   Butylene (C_4H_8)

Fig. 5-7—Olefins

For example, sticking a lot of ethylenes to each other makes polyethylene.

The key characteristic of the olefins is the absence of two hydrogens from an otherwise *saturated* paraffin—*i.e.*, a paraffin that has a full complement of hydrogens. The formula for an olefin, therefore, is C_nH_{2n}.

Olefins are unnatural. It may be that only God can make a tree, but God could not handle olefins. They are not found in crude oil but are man-made in one of the several cracking processes covered in later chapters. This seemingly irrelevant fact has implications in the design of refinery gas plants, which generally keep saturated gas and cracked gas segregated. More on this in chapter 7.

The other olefins that will be of primary interest in petroleum refining will be propylene (C_3H_6) and butylene (C_4H_8). Like ethylene, these compounds can be reacted with other chemicals easily and so are suitable for a number of both petrochemical and refinery applications.

Aromatics are another exception to the valence rules. Take the ring compound cyclohexane. Each carbon has two hydrogens attached and is attached to two other carbons. If you removed one hydrogen from each carbon, you could bring the valence rules back to being satisfied by putting in some double bonds between the carbons. A double bond between *every other* carbon will do it (Fig. 5-8). The resulting molecule is C_6H_6, *benzene*. Saying that every other bond around the benzene ring is a double

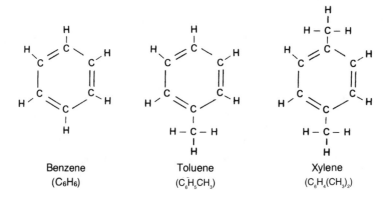

Benzene
(C_6H_6)

Toluene
($C_6H_5CH_3$)

Xylene
($C_6H_4(CH_3)_2$)

Fig. 5-8—Aromatics

bond is a gross simplification. The real description has to do with a resonating structure and hopping bonds, but that's another story.

If one of the hydrogen of the benzene molecule is removed and a methyl radical, $-CH_3$, is put in its place, the result is $C_6H_5CH_3$, toluene. If two methyl radicals are substituted for two, not one, hydrogen on the benzene ring, the result is $C_6H_4(CH_3)_2$, xylene (pronounced zi-leen).

As with ethylene, the double bonds make the benzene ring very unstable and chemically reactive. That makes benzene a very popular building block in the chemical industry. Toluene and xylene are also important chemical feedstocks. The instability of the double bonds on the aromatics ring is best illustrated by the violent chemical reaction of the most notorious of the aromatics derivatives, tri-nitro-toluene (TNT).

The aromatics include *any compounds* that have the benzene ring in them. At the same time, many people use the term aromatics to refer to the BTXs, the benzene, toluene, and xylenes. The name aromatics, by the way, comes from the characteristic smell of the BTXs, which is a sickly sweet hydrocarbonish odor. Chapter 16 deals with the special apparatus for the extraction of aromatics in refineries.

You can see that once the number of carbons gets above six, the number of different variations in structure explodes. For that reason, refiners pay little attention to the individual compounds above the C_6's. Maybe a few C_7's or C_8's, but that's about it. Sometimes they will use the *PONA's*, the proportion of paraffins, olefins, naphthenes, and aromatics, to evaluate the suitability of a stream; other times they will use the physical properties (gravity, viscosity, boiling temperatures, etc.) as an expedient.

Exercises

1. Isobutane is the isomer of normal butane. Why isn't there an isomer of propane?

2. How many different ways can isobutane be structured? Isopentane? Isobutylene?

3. Name the four types of structures generally referred to as the PONAs.

4. There are three different kinds of xylene. Can you draw the two not shown in Figure 5-8?

5. Why is there only one type of toluene?

chapter six
6
Cat Cracking

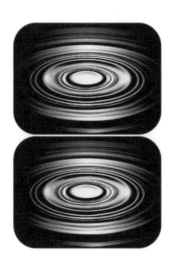

What cracker is this that deafs our ears
With this abundance of superfluous breath?

-*King John*, William Shakespeare

I n the adolescent years of the petroleum industry, the proportion of the crude oil barrel that consumers wanted in the form of gasoline increased faster than fuel oil. It became apparent to refiners that distilling enough crude to make straight run gasoline to satisfy the market would result in a glut of fuel oils. That would reverse the situation in the 19th century when they literally dumped gasoline on the ground as they processed enough crude oil to make the fuel oils. The new economy saw increasing prices of gasoline and declining prices of the heavier cuts.

To cope with this physical and economic problem, inventive process engineers developed a number of cracking techniques, the most popular of which eventually became *cat(catalytic) cracking*.

The Process

In the chapters on distilling and vacuum flashing you read about the phenomenon of cracking and you saw a little about the chemistry of cracking in Figures 4-1 and 4-2. Here's the process of cat cracking: in a cat cracker, straight-run heavy gas oils are subjected to heat and pressure and are contacted with a *catalyst* to promote cracking.

Definition: a catalyst is a substance added to a chemical reaction that facilitates or causes the reaction but when the reaction is complete the catalyst comes out just like it went in. In other words, the catalyst does not change chemically. It causes reactions between other chemicals.

Catalysts are like some red-headed 10-year-olds you know. They never get into trouble—it's just that wherever they go, trouble happens.

Catalysts can be liquids, gelatins, or solids, and those solids can be powders, beads, or shaped pieces.

The feed to the cat cracking process is usually straight-run heavy gas oil and flasher tops. The boiling range cat feed can be anywhere in the 650-1,100°F range. Heat is required to make the process go; temperatures in the cat cracker where the cracking takes place can be about 900°F or higher, even approaching 1,100°F depending on the catalyst and the results desired.

The process is designed to promote cracking in a specific way. The object is to convert the heavy cuts to gasoline. Ideally, the entire product coming out of the cat cracker would be in the gasoline range, but the technology is not that good. During the cracking process, lots of phenomena occur:

- As the large molecules crack, there isn't enough hydrogen to go around so some small amounts of carbon form *coke*, which is virtually pure carbon atoms stuck together.
- As the large molecules break up, a full range of smaller molecules from methane on up are formed. Due to the deficiency of hydrogen, many of the molecules are olefins.
- Where the large molecules in the feed are made up of several aromatic or naphthene rings stuck together, smaller aromatic or naphthenic compounds plus some olefins result.
- Finally, the large molecules, made up of several aromatic or naphthenic rings plus long side chains, are likely to crack where the side chains are attached. The resulting molecules, though lower in carbon count, are more dense or heavier; their specific gravity is higher, their API gravity is lower. They also tend to have higher boiling temperatures. Ironically these molecules form a product heavier than the feed.
- The products of cat cracking are the full range of hydrocarbons, from methane through to residue, plus coke.

Three main parts make up the cat cracking hardware—the reaction section, the regenerator, and the fractionator.

Reaction Section

The guts of the cat cracker is the reaction section (Fig. 6-1), which consists of the cat feed heater and the riser (a pipe from ground level up to a water tank-looking vessel called a disengagement chamber). The heater

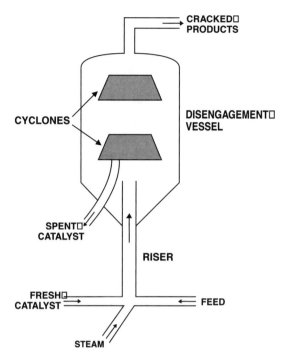

*Fig. 6-1—*Cat Cracker Reaction Section

raises the temperature of the cat feed to 900-1000°F; the feed mixes with catalyst being pumped into the riser. Steam is introduced with the catalyst to give the whole mixture enough lift to climb up to the bottom section of the disengagement chamber. Because of the temperature and the intimate contact with the catalyst, all the reactions listed above take place in the riser, even though *residence time* in the riser is only seconds. In the older cat crackers, the disengagement chamber was called the *reactor* because that's where most of the chemical changes took place. The newer crackers use this vessel only to separate the catalyst from the hydrocarbon.

As the hydrocarbon/catalyst mixture hits the disengagement chamber, it encounters a *cyclone*, a mechanical device that spins the mixture. The catalyst being heavier, the centrifugal motion slaps it against the walls of the cyclone where it slides to the bottom and out the piping, exiting the disengagement chamber via gravity. The hydrocarbon, mostly in vapor form, but with some liquid droplets, rises from the cyclone to the top of the chamber, encounters another cyclone that does the final clean-up act, then exits the top.

Refiners had a purpose in switching to a design where the reactions take place in the riser instead of in the reactor—to lengthen the contact time between the feed and the catalyst. However, on some *riser crackers,* as the new vintage of catalytic cracking units (CCUs) is sometimes called, the hardware is set up to allow some feed to be introduced further up the riser to reduce the contact time. That allows refiners to segregate the feed from different crudes, since feeds with different compositions often respond to the catalysts at different rates.

Catalysts

The catalyst used in modern cat crackers is a marvel of evolution. It used to be made of natural alumina-based clay but now refiners buy only the much-improved synthetically produced cat cracking catalysts called zeolytes. The particles have three unusual characteristics. If you had a jar of cat cracker catalyst and shook or tilted it, the powder, which looks like off-white baby powder, would slosh around just like a fluid. This behavior, so important to the design of the whole process, gave rise to another name refiners often use, *fluid cat cracking.*

The second characteristic is not apparent to the naked eye. Under a microscope you would be able to see that each catalyst particle has a large number of pores, and as a consequence, a tremendous *surface area,* especially in relation to the size of the particle. If the particles were the size of the planet earth, the pores would be deeper than the Grand Canyon and spaced every few miles or so over the surface. The influence of the catalyst depends on contact with the cat feed so the huge surface area is vital to the process.

The third characteristic in the modern cat cracker catalysts comes from technology leaps. The old alumina-based clay catalyst had the minerals necessary to promote the cracking reactions. Nowadays the catalysts are synthesized to exacting dimension and mineral content. The pores are designed and fabricated so minutely that they will let in just one molecule at a time, and only molecules of a certain size. That way the types of molecules that are catalyzed can be controlled and refiners can end up with designer outturns. Well, almost. The catalyst suppliers can furnish refiners with catalysts that will favor (but not deliver 100%) the creation of high octane gasoline components or, perhaps, light olefins, or permit the use of heavier feeds; create less coke; reduce the temperature of the reaction to save energy; and so on.

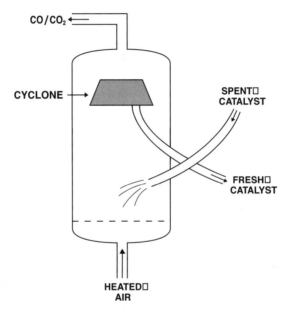

Fig. 6-2—Catalyst Regenerator

The Regenerator

During the cracking process, some portion of hydrocarbon cracks all the way to coke and ends as a deposit on the catalyst. As the catalyst surface is covered, the catalyst becomes inactive (*spent*) and reduces its effectiveness. To remove the carbon, the spent catalyst flows by gravity to a vessel called a regenerator (Fig. 6-2). Heated air, about 1,100°F, is mixed with the spent catalyst and a chemical reaction takes place:

$$C + O_2 \longrightarrow CO \text{ and } CO_2 \text{ (older cat crackers)}$$
$$C + O_2 \longrightarrow CO_2 \text{ (newer cat crackers)}$$

This process, *oxidation of coke*, is similar to burning coal or briquettes in that carbon unites with oxygen and gives off carbon dioxide (CO_2), perhaps carbon monoxide (CO), and a large amount of heat. The heat, in the form of the hot CO/CO_2, is generally used in some other part of the process, such as raising steam to drive pumps or turbines. In the older cat crackers, the CO/CO_2 is sent to a CO furnace where oxidation of the rest of the CO to CO_2 is completed before the CO_2 is finally blown out to the atmosphere.

The regenerator has its own cyclone at the top that separates the catalyst from the CO/CO_2. The regenerated catalyst flows from the cyclone, again by gravity, out of the regenerator, ready to be mixed with cat feed and steam and sent up the riser once more. Thus the catalyst is in continuous motion going through the cracking/regeneration cycle.

The Fractionator

Meanwhile, back on the hydrocarbon side, when the cracked product leaves the reaction chamber, it is charged (pumped in) to a fractionating column dedicated to the cat cracker product. The products separated generally are the gases (C_4 and lighter), cat cracked gasoline, cat cracked light gas oil, cat cracked heavy gas oil, and the fractionator bottoms, called cycle oil. A variety of things can be done with the cycle oil, but the most popular is to mix it with the fresh cat feed and run it through the reaction again. Some of the cycle oil cracks each time through the reactor. By recycling enough, all the cycle oil can be made to disappear. The process has the ominous designation *recycling to extinction*. Sometimes the most stubborn molecules just keep going around in a circle with no further cracking, so a small amount of cycle oil can be drawn off continuously and blended off to heavy fuel oil.

The cat cracked heavy gas oil can be used as feed to a hydrocracker or thermal cracker or as a residual fuel component. The light gas oil makes a good blending stock for distillate fuel, and the cat cracked gasoline a good motor gasoline blending component.

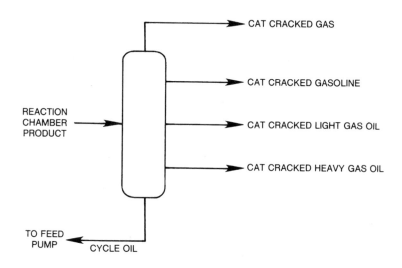

Fig. 6-3—**Fractionation**

There is quite a bit of latitude in the cut point between the gasoline and light gas oil stream. Refiners use this as one way to regulate the balance between gasoline and distillate as the seasons change. As the winter heating oil season arrives, many refineries go into a *max distillate mode*. They make adjustments to the CCU fractionator to lower the end point of the cat cracked gasoline to push more volume into the cat light gas oil. In the summer, during a *max gasoline mode*, the shift is in the other direction.

The products from the fractionator are different in composition than those from the crude distilling column light ends. The cracking process results in the creation of olefins, so the C_4 and lighter stream not only contain methane, ethane, propane, and butanes but also hydrogen, ethylene, propylene, and butylenes. Because of these extra components, this stream is sent to be separated at the *cracked gas plant*. This is in contrast to the gas from operations like distilling (and, as discussed later, the hydrotreater, hydrocracker, reformer, and others) where the gases contain only saturated compounds. These end in the *sats gas plant* for separation. The isobutane, propylene, and butylenes from the cat cracker will be of special interest when discussing alkylation, a process that converts these compounds into gasoline blending components.

The other heavier products also differ in composition. During the cracking process, many of the heavy, complex molecules will crack at the connection between the aromatics rings and the side chains. Consequently, the cat cracker products tend to be rich in aromatics, the molecules replete with the benzene ring somewhere in the structure. That's good news for boosting octane numbers and making gasoline, but as you'll find out later, bad news for making jet fuel and diesel fuel.

All pieces fit together as shown in Figure 6-4. You can see there are two circular flows going on. On the left side, the catalyst goes through the reaction, is regenerated and gets charged back to the reaction again. On the right side, hydrocarbon comes in and goes out, but the cycle oil provide continuous circulation of at least some of the hydrocarbon components.

Yields

The object of the cat cracking process is to convert heavy gas oil to gasoline and lighter. A set of typical yields will demonstrate how successful the process is.

		% Volume
Feed:	Heavy Gas Oil	40.0
	Flasher Tops	60.0
	Cycle Oil	(10.0)*
		100.0
Yield:	Coke	8.0
	C_4 and Lighter	35.0
	Cat Cracked Gasoline	55.0
	Cat Cracked Light Gas Oil	12.0
	Cat Cracked Heavy Gas Oil	8.0
	Cycle Oil	(10.0)*
		118.0

*Recycle stream not included in feed or yields total.

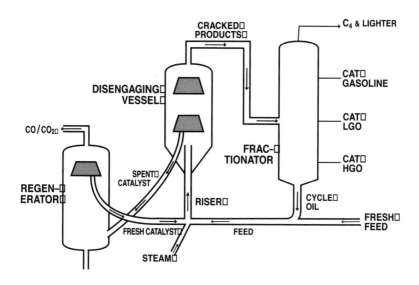

Fig. 6-4—**Cat Cracking Unit**

Since the cycle oil is recycled to extinction, a simple diagram for cat cracking doesn't even show it as entering or leaving the process. Something more important does show up—however, the phenomenon of *gain*. In the yield structure shown above, the products coming out add up to 118% of the feed going in. That has nothing to do with the recycle stream but only with the gravities of the products coming out compared to the gravities of the feeds. If you measured the yields in percent weight instead of percent volume, the yields would come out to 100. But since most US petroleum products are sold by the gallon, US refiners measure everything in volume. Since cracking plays games with the densities, cat cracking yields show a substantial gain. Gain is the bane of the accountants but sometimes becomes an obsession with refiners as they attempt to "fluff up the barrel" with their cracking processes.

Process Variables

Cat crackers usually make so much money by converting heavy feed to light products that refiners almost always run them to their capacity. The limit could be set by *coke burning capacity*, (the rate at which the catalyst is regenerated), by feed rate, or by fractionator capacity if the reaction makes too much of one of the products. The bottlenecks become apparent when the yields of gasoline fall off and either heavy gas oil or C_4 and lighter yields start to increase. Some of the things that affect cat cracker yields are the quality of the feed, the reaction temperatures, the feed rate, the recycle rate, and, incredibly enough, the time of day and temperature outside the control room, as explained below.

Feed quality. The cracking reaction is very complex, and lots of data are available to predict yields from feed with different characteristics. The gravity of the feed and the paraffin, naphthene, and aromatic content are important indicators.

Reaction temperature. The higher the temperature, the more cracking will occur; but at some point, the generation of C_4 and lighter gases will really take off, at the expense of the cat gasoline or cat light gas oil.

Feed rate and recycle rate. The yields will suffer at higher feed rates since the contact with the catalyst will be diluted. So refiners watch the *fresh feed rate* and the volume of fractionator bottoms being recycled.

Time of day and temperature. Cat crackers have better yields at night than the day. Are the swing shift better operators? Do tinkering engineers come around only during the day and mess things up? Not really. In order to regenerate the spent catalyst, fresh air is heated and pumped into the regenerator continuously. As the temperature of the air outside goes down, the air gets denser. Since the blowers that pump the air are always operated at maximum speed, more oxygen is actually pumped into the regenerator when it's cold than when it's hot. The more oxygen, the more coke burned off the catalyst. The fresher the catalyst, the better the reaction. The better the reaction, the more gasoline produced. From the operator's data logs, the swings can actually be plotted, and as night comes and the temperature drops, the yield gets better. With the heat of the afternoon, the yield falls off. Summer and winter have the same effects, which is too bad because usually the demands for gasoline go up in the summer when the yields go down.

Review

From a process point of view, you can sketch the cat cracker as a feed-in/product-out box in a refinery flow diagram. Thus far, you've covered distilling, flashing, and cat cracking. Figure 6-5 shows the way these come together. This diagram introduces the shorthand notation that will be used in building the entire refinery flow diagram from here on.

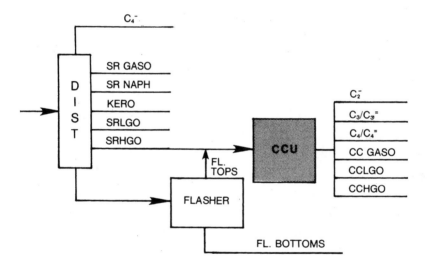

Fig.6-5—**Refinery Flow Diagram**

Exercises

1. What's the difference between a cat cracker, a fluid cat cracker, and a riser cracker?

2. Fill in the blanks:
    a. There are two circulating flows in a fluid cat cracker. On the one side is _____; on the other is _____.
    b. Spent catalyst has _____ deposited on it. The regeneration process removes it by reacting it with _____ to form _____ and _____.
    c. The purpose of cat cracking is to convert _____ to _____.
    d. Cat cracker feed usually comes from the _____ and the _____.
    e. _____ are found in the cracked gas stream but not in the sats gas streams.
    f. The cat-cracker fractionator bottoms, called _____ _____, are usually _____ _____ _____ by injecting them in the feed.

3. Use the data and answers from the previous chapters and the cat cracker yields from this chapter. Assume feeds to the cat cracker are all the flasher tops and all the straight-run heavy gas oil (the cut between the straight-run light gas oil and flasher feed). How much cat light gas oil is produced?

4. An indicator of how complicated refining operations can get is the number of alternatives to make a downstream volume change. Name six ways to increase the volume of cat light gas oil.

c h a p t e r   s e v e n

Refinery
Gas Plants

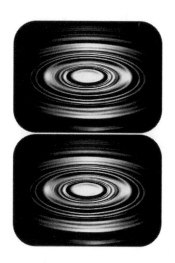

...and he shall separate them one from another,
as a shepherd divideth his sheep from the goats.

-Matthew 25:32

Almost all refinery processing units generate some volumes of butane and lighter gas. Gas processing is not a very colorful operation, but maybe that's why the engineers gave such exotic names to the parts. There's lean oil, fat oil, sponge oil, and rectified absorbers, to mention a few. All are in the central location called the gas plant.

Sats Gas Plant

The gas streams from most of the refinery process units have only *saturates* in them: methane, ethane, propane, butanes, and maybe hydrogen. The word *saturates* is a synonym for hydrocarbons with no double bonds; *i.e.*, all the carbon atoms are "saturated with hydrogen atoms." These streams are usually handled in a facility called the *sats gas plant*. In contrast the *cracked gas plant* handles the streams that also contain the olefins (ethylene, propylene, and butylenes). Dealing with the even lighter gases, hydrogen and hydrogen sulfide is so special that you have a separate chapter devoted to it, chapter 15.

Some Properties of Light Ends

Compound	Formula	Boiling Temperature* (°F)	Density** (lbs/gallon)
Hydrogen	H_2	−423	—
Methane	C_1H_4	−258	2.5
Ethane	C_2H_6	−127	2.97
Ethylene	C_2H_4	−155	3.11
Propane	C_3H_8	− 44	4.23
Propylene	C_3H_6	− 54	4.37
Isobutane	C_4H_{10}	11	4.69
Normal Butane	C_4H_{10}	31	4.87
Isobutylene	C_4H_8	20	5.01
Normal Butylenes	C_4H_8***	21	5.01
	C_4H_8	34	5.09
	C_4H_8	39	5.23

*At atmospheric pressure
**At atmospheric pressure and 60° F vacuum
***There are three forms of normal butylenes, each of them with the formula C_4H_8;
  each with a unique structure; each with unique properties

Table 7-1—Some Properties of Light Ends

The separation of the gases is a lot harder than the separation in the crude distilling unit. Remember that each of those gases is a single chemical and boils at a single temperature (Table 7-1). That doesn't leave much tolerance for tail ends or any kind of sloppy separation. As a result, the gas separation columns have lots of trays and lots of reboiling and refluxing.

Another complication is the pressure/temperature relationships. The gas plants are just the inverse of the flasher. In order to get the streams to liquefy—an essential step in distillation—the mixture has to be either supercooled (see table on previous page) or put under a lot of pressure. Gas separation usually uses both in combination.

The saturated or *sats* gas coming from the processing units to the gas plant can go through the following typical steps:

1. *Compression* and *phase separation*. Low pressure gases are compressed to about 200 psi, as in Figure 7-1. Because of the high pressure, some of the gases liquefy, allowing them to be drawn off.

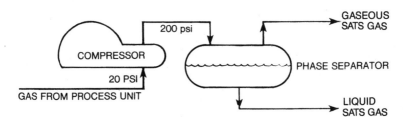

Fig.7-1—Compression and Phase Separation

2. *Absorption*. The remaining high pressure gases are introduced to the lower section of the *rectified absorber* (Fig. 7-2). This unusually shaped column is a fractionator with a special counterflow built in. *Lean oil*, a naphtha range liquid, enters the top tray of the absorber and trickles down the column. As it does, it picks up (absorbs) most of the remaining propane and butane in the gas. Continuing the gastronomic analogy, the naphtha laden with the propane and butane is called *fat oil*. The rocket ship design of the column has to do with the masses and velocities of the gases and the naphtha as they pass each other, with more propane and

butane being in the mixture at the bottom and less by the time the gas is scrubbed at the top.

3.   *Debutanizing.* The fat oil is charged to a column, called a debutanizer (Fig. 7-3) that separates the butanes and propane from the naphtha. Because the boiling range of the naphtha starts about 180°F and butane boils at about 31°F, the split is relatively easy. It's certainly easier than splitting butane and propane from ethane in a distilling column. The liquid portion from the phase separator is charged to the debutanizer also

4.   *Depropanizing.* The C_3 /C_4 stream is separated in a tall column that takes the propane overhead.

5.   *Deisobutanizing.* Some refineries have a column that splits the butane. Since the boiling temperatures of normal and isobutane are so close, many trays are needed to get good separation. Lots of chillers and compressors are needed to liquefy the gas and reintroduce it as a reflux. The deisobutanizer is usually the tallest column in the gas plant.

6.   *Sponge absorption.* One complication occurs when the lean oil is introduced to the rectified absorber. Some of the lean oil, usually heptane (C_7H_{16}) or octane, vaporizes and goes out the overhead with the ethane and lighter (C_2^-). To recover the lean oil, the overhead is charged to the bot-

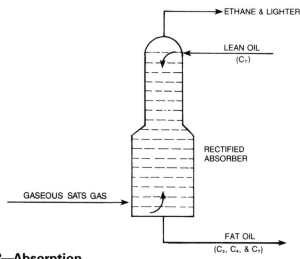

Fig. 7-2—Absorption

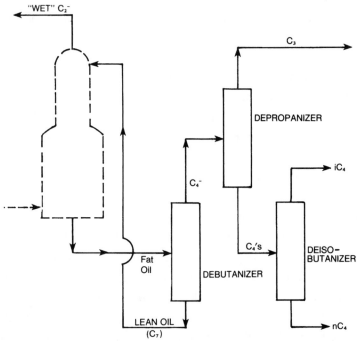

***Fig. 7-3*—Debutanizing, Depropanizing, Deisobutanizing**

tom of another absorber column and another heavier lean oil, called *sponge oil*, charged to the top. The overhead is almost all C_2^-, and the bottoms are almost all lean oil as shown in Figure 7-4. Sponge absorption is very similar to rectified absorption, and sponge oil is almost the same as lean oil, but the names are so colorful you'll want to be able to toss all of them around.

Ethane is usually not split out in a refinery, since its primary use is refinery fuel. Sometimes, however, the ethane might be needed for feed to a chemical plant. Then the C_2^- is de-methanized in another absorber column, separating the ethane and methane.

Where the propane market is strong, refiners sometimes want to squeeze the last bits of propane out of the C_2^- stream, the volume that would otherwise be considered tail ends. In that case they might put the C_2^- stream through another chiller to drop out more liquid propane to add to the earlier propane stream.

All the main steps are put together in Figure 7-5.

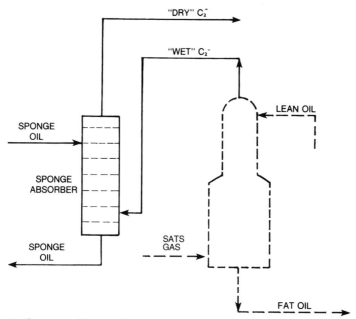

Fig.7-4—Sponge Absorption

Disposition

Why do refiners separate all these streams? Each of the different components has found a unique home in the refining business, and the highest value can usually be attained only when the stream is separated for disposition to this use.

Isobutane. Used predominantly as one of the feedstocks to the alkylation process. Occasionally used as a motor gasoline blending component. Sometimes as feed to a dehydrogenation unit on the way to making MTBE. (If this makes you curious, see chapters 16 and 17.)

Normal butane. Used predominantly as a motor gasoline blending component. Due to its volatility, it is good for starting cold engines but the rate at which it evaporates limits the amount that can be added. Some normal butane is also used as a chemical feedstock and as LPG mixed with propane, and as feed to a butane isomerization plant where it is converted to isobutane.

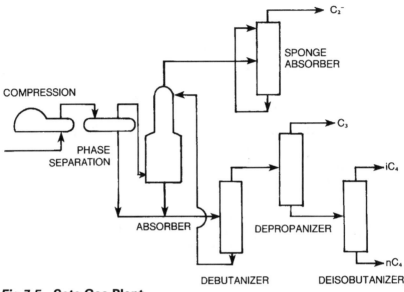

Fig. 7-5—Sats Gas Plant

Propane. In the US most LPG (liquefied petroleum gas) is propane. About 10% of the LPG sales are butane or butane-propane mixes. Propane as a fuel has some unique characteristics that have found favor in many applications. It can be liquefied at reasonable temperatures and pressures for ease of transport, particularly in trucks. Yet at ambient temperatures, it will readily vaporize, making it easy to burn in stoves and home furnaces with no fuel pump and therefore no electricity. Propane has also been widely used as a chemical feedstock for making ethylene and propylene.

Ethane. The only application that gives rise to separating ethane in a refinery is its use as a chemical feedstock for making ethylene (chapter 18, "Ethylene Plants"). Otherwise, the ethane is left with the methane.

Methane. Refineries have a voracious appetite for fuel. Methane finds a ready home in all the process furnaces and steam boiler furnaces around the refinery.

Hydrogen. Several refinery processes require hydrogen—hydrocracking and a variety of hydrotreating units. Some of the gas streams, in particular the stream from the reformer, have high enough concentrations of hydrogen to warrant keeping them segregated through the compression stage. Compression then permits separation of a *hydrogen concentrate*, an H_2/C_1 mix that can be used in the hydrogen applications. A supplementary supply of hydrogen can be specially made if the refinery has a *hydrogen plant*, sometimes called a *steam-methane reformer*, which is a subject in chapter 15.

Cracked Gas Plant

The gas streams from the cat cracker and the other cracking processes contain olefins—ethylene, propylene, and butylenes. Keeping the cracked gases apart from the saturated gases eliminates the extra fractionating that would be required if they were commingled.

Usually ethylene stays with the C_2^- stream going to the *refinery fuel system*. It could be separated if there were a need for it in the chemical plant. But the propane/propylene and the butane/butylene streams are separated from the C_2^- stream and from each other in fractionators much like those in the sats gas plant.

In a refinery, the propylene and butylenes have traditionally been sent to an *alkylation plant*. For convenience the saturates, propane and butane, are sent along as well and separated there (chapter 8). Sometimes the propane/propylene stream or the butane/butylene stream is sent off to a chemical plant as feedstock. The advent of MTBE plants in refineries also led to separating iso-butylene, one of the two MTBE feedstocks, from the other C_4's.

Storage Facilities

This is a good place to talk about the rather specialized facilities needed to store the refinery gases. Because of the volatility and boiling temperatures of these streams, high pressure containers are needed to store the streams in the liquid form.

Methane (and ethane not used as chemical feedstock) is usually never stored in a refinery but is sent directly to the fuel system as it is produced. There are some *surge drums* in the system that can accumulate some of the gaseous methane for short periods of time during operating changes.

Also those tall *flares* so characteristic of refineries are used to handle momentary surges in gas manufacture by burning the excess.

Propane and butanes (and sometimes ethane) can be stored in steel storage or in underground storage. The characteristic of the propane and butane storage is the spherical shape. Steel storage is either in the form of *bullets* (cylinders on their sides with rounded ends) or large spheres. The round shape is due to an optimization of structural strength to accommodate the high pressure and the cost of the steel.

Underground storage generally takes one of two forms—*caverns* mined in rock, shale, or limestone, or *jugs* leached out of salt in underground *salt domes* as shown in Figure 7-6. In the case of mined storage, the propane or butane is pumped in and out of the cavern in the liquid form,

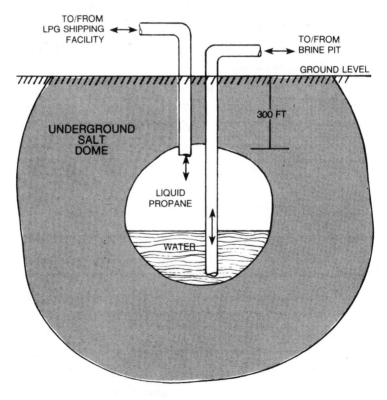

Fig.7-6—Underground Storage Jug for JPG

with the difference in the amount contained made up by more or less vapor being formed above the liquid.

In a salt dome, the ins and outs are handled in a more unusual fashion. The jug contains a combination of propane and salt water (brine). The two liquids, like any other oil/water combination, do not mix. To get the propane out, more brine is pumped in, forcing the propane out. To fill the jug, propane is pumped in, forcing the brine out, usually to a brine pit on the surface.

One advantage of salt dome storage is its expandability. By pumping fresh water instead of brine, more of the salt from the jug walls is dissolved, expanding the size of the jug at no cost. There is, of course, a practical limit to the size—the jug could get so wide it could collapse or breach the integrity of the salt dome and leak out the edges.

The other advantage is the original construction cost. Salt dome jugs are much cheaper than mined storage, which is much cheaper than steel storage. Unfortunately, nature limits the availability of salt domes, suitable rock, shale, and limestone to a few areas around the country.

Exercises

1. Fill in the blanks:
   a. The two sections of the gas plant are called the _____ gas plant and the _____ gas plant.
   b. The naphtha used to strip out the heavier gases in the rectified absorber is called _____ _____. The liquids-laden lean oil is called _____ _____.
   c. A stream that can be used to absorb a lot of some parts of another stream is called a _____ _____.

2. Match up the following streams with the usual uses:

Streams	Used in
Methane	
Ethane	Hydrotreating
Propane	Blended to motor gasoline
Normal Butane	Refinery fuel
Isobutane	Alkylation feed
Propylene	Chemical Feed
Butylenes	Commercial fuel
Ethylene	
Hydrogen	

3. Draw a set of distillation columns that could be used to separate the cat cracker gases into C_2 and lighter, propane, propylene, normal butane, isobutane, and butylenes. Use Table 7-1 that has the boiling temperatures of each compound. Assume each column has only tops and bottoms, no side draws. What special problem does the butane/butylenes split present?

4. Make a flow diagram of the streams in and out of the units already covered, the crude distilling column, the flasher, the cat cracker, and the gas plant.

c h **a** p t e r   e i g h t

Alkylation

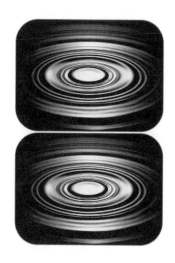

Yield who will to their separation,
My object in living is to unite.

-Two Tramps in Mud Time
Robert Frost

After the engineers were so clever about the invention of cat cracking, they attacked the problem of all the light ends the process created. The objective was to maximize the volume of gasoline being produced, but butylenes and propylene were too volatile and plentiful to stay dissolved in the gasoline blends. So they devised a process that was the inverse of cracking, *alkylation*, which starts with small molecules and ends up with larger ones.

The Chemical Reaction

To a chemist, alkylation can cover a broad range of reactions that stick molecules together. To a refiner, alkylation is the reaction of propylene or butylene with isobutane to form an iso-paraffin called alkylate (Fig. 8-1).

Alkylation has a volumetric effect on refining operations the inverse of cracking because there is a significant amount of *shrinkage*. With propy-

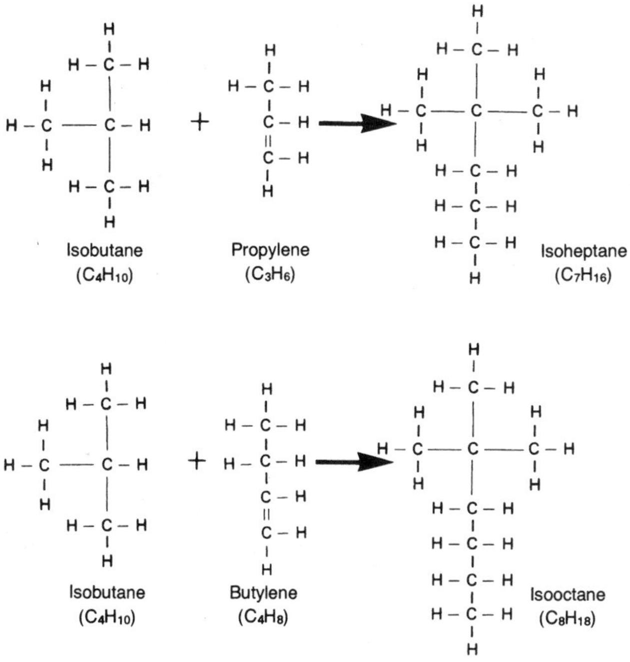

Fig.8-1—Alkylation of Propylene and Butylene

lene as the feed, 1 bbl of propylene and 1.6 bbl of isobutane go in and 2.1 bbl of product come out; 1 bbl of butylene and 1.21 bbl of isobutane yield 1.8 bbl of product. As in cracking, the weight in equals the weight out. Nothing gets lost. Only the densities and volumes change.

The Process

Propylene and butylenes are hyper enough that the chemical reaction could be made to take place by just subjecting the isobutane and olefins to high pressures. However the equipment would be very expensive to handle this route to alkylation. Like a lot of other processes, catalysts have been developed to facilitate the process and simplify the hardware. Alkylation plants use either sulfuric acid or hydrofluoric acid as the catalyst. In this case the catalyst is liquid, in contrast to the solids in cat cracking. The processes for both are basically the same. Those plants that use hydrofluoric acid are called HF plants; the others are called sulfuric plants. Both have safety concerns because hydrofluoric and sulfuric acids are seriously nasty items, corroding all but the specially lined vessels and piping around them. Hydrofluoric acid has an additional concern. If it escapes, it floats in a cloud and can travel great distances to the annoyance of refinery neighbors. Sulfuric acid will form droplets if it escapes and quickly settles down to the ground, though that's not much consolation to anyone working in the immediate vicinity.

Sulfuric plants work better for butylene alkylation; HF plants are better for propylene. Sulfuric acid seems to be slightly more popular, so this chapter describes only the sulfuric route. However, the HF route to alkylation is not much different.

The *alky plant* consists of seven main parts: the chiller, the reactors, the acid separator, the caustic wash, and three distilling columns, all in Figure 8-2.

The chiller. Alkylation with sulfuric acid catalysts works best at temperatures in the neighborhood of about 40°F. The olefin feed (a propane/ propylene or butane/butylene stream from the cracked gas plant) and a stream of isobutane is pumped through a chiller and mixed with a stream

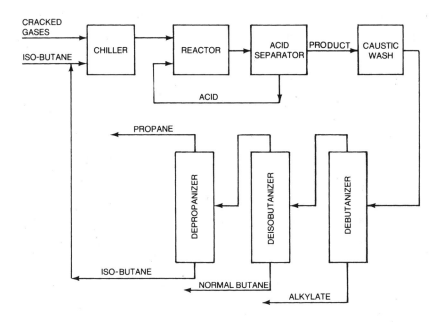

Fig.8-2—**Alkylation Plant**

of sulfuric acid. The pressure is high enough to keep the mixture in liquid form. Sometimes the chilling is done right in the reactor.

The reactors. The reaction time for the alkylation process is relatively long, so the mixture is pumped into a battery of large reactors. The reactors hold so much total volume that by the time they all turn over once, the *residence time* of any one molecule is quite long, about 15 to 20 minutes. As the liquid passes through the reactors, it encounters mixers to assure that the olefins come in good contact with the isobutane and the acid catalyst, permitting the reaction to occur.

Acid separator. The mixture then moves to a vessel where no mixing takes place, and the acid and hydrocarbons separate like oil and water. The hydrocarbon is drawn off the top; the acid is drawn off the bottom. The acid is then recycled back to the feed side. The acid separator is also referred to as the *acid settler*.

Caustic wash. The hydrocarbon from the acid separator will have some traces of acid in it, so it is treated with caustic soda in a vessel. Caustic soda does to the hydrocarbon what Alka-Seltzer® does to your stomach when you have indigestion—it neutralizes the acid. What's left (in the alky plant, not your stomach) is a mixture of hydrocarbons ready to be separated.

Fractionators. Three standard fractionators separate the alkylate and the saturated gas. Any unreacted isobutane is recycled to the feed.

Yields

During the alkylation process, a number of *side reactions* occur, some of which are more or less undesirable. Because there are a lot of molecules forming and reacting, there are small amounts of propane, butane, and pentane formed, which are not too bad, but a small amount of *tar* forms as well—a thick brownish oil containing a mixture of complex hydrocarbons. The molecules are so heavy they usually build up in the acid and exit the scene when the acid is sent back to the supplier for regeneration.

The volumetric balances for propylene and butylene feed are shown in Table 8-1:

	Volume Balances	
Feeds	Propylene	Butylene
Propylene	1.0	—
Butylene	—	1.0
Isobutane	1.6	1.2
	2.6	2.2
Products		
Propane	0.3	
Normal Butane	—	0.1
Alkylate	1.8	1.7
	2.1	1.8

Table 8-1—Alkylation Yields

What the feeds and the yields don't show is the propane and normal butane that just pass through the plant untouched. As a matter of fact, the alky plant provides significant capacity to the refinery for making segregated propane and butane streams. The reactors take out the propylene and butylene, converting it to alkylate, and the fractionators separate the propane and butane from the alkylate. When the alky plant is down for one reason or another, the butane/butylene stream might find another home in gasoline blending. The propane/propylene stream is too volatile to put in gasoline so it is usually routed to the refinery fuel system instead and LPG production (propane only) drops significantly.

Process Variables

The alky plant manager has to watch a number of key variables to keep too many side reactions from occurring that could cause the quality of the alkylate to deteriorate, as evidenced by such things as lower octane number, poor color, and high vapor pressure.

Reaction temperature. Temperatures too low cause the sulfuric acid to get syrupy. That inhibits complete mixing and the olefins do not completely react. High temperatures cause compounds other than isoheptane and isooctane to occur, lowering the alkylate quality.

Acid strength. As the acid circulates through the process, it gets diluted with water that inevitably comes in with the olefins and also picks up tar. As the acid concentration goes from 99% down to about 89%, it is drawn off and sent back to the acid supplier for refortifying.

Isobutane concentration. By having an excess amount of isobutane, the process works better. Isobutane recirculation systems are generally built in. The ratio of isobutane to olefin varies from 5:1 to 15:1. The room in the reactors usually limits the concentration.

Olefin space velocity. The length of time the fresh olefin feed resides in the reactor causes alkylate quality to vary.

Review

Alkylate has emerged as a hero in the past few years as refiners struggle to improve the environmental qualities of gasoline. Since you haven't gotten to chapter 12 on gasoline blending, you will not fully appreciate the attributes of alkylate. It has a low vapor pressure, zero sulfur, zero olefin content, zero benzene, and a high octane number—a blender's dream. All this comes from working with some otherwise low-valued cats and dogs around the refinery—propylene, butylene, and iso-butane.

In the big refining picture the alky plant can be represented by a box with propylene, butylene, isobutane, propane, and normal butane inside. On the outside are the alkylate, propane, and normal butane. To put alkylation in perspective, Figure 8-3 shows the refinery processing units covered so far, plus alkylation. How complicated it quickly gets.

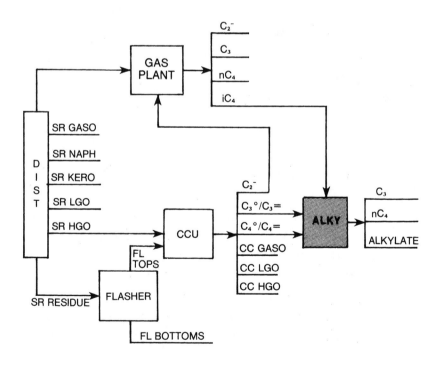

***Fig.8-3*—Refinery Processing Operations**

Exercises

1. Fill in the blanks.

   a. Alkylation is the inverse of _____.

   b. The catalyst in alkylation is either _____ or _____, depending on the process chosen.

   c. The five main parts of an alky plant are:

   d. Alkylate is made up mostly of two isoparaffins called _____ and _____.

   e. Because of the high _____ _____, low _____, and zero _____, _____, and _____ content, alkylate makes a good _____ _____ _____.

2. The cracked gas stream coming to an alky plant has the following composition.

	Volume
Propane	.15
Propylene	.25
Isobutane	.10
Normal butane	.20
Butylene	.30
	1.00

   If the volume is 3,000 B/D and the alky plant yields are those shown in this chapter, how much isobutane is needed from the sats gas plant? What are the alky plant outturns?

9

Catalytic Reforming

Nothing so needs reforming
as other people's habits.

-Pudd'nhead Wilson,
Mark Twain

C at reforming has provided more controversy in the refining business than all the other units combined. What started out as an engineering solution to market needs ended in the middle of debates between refiners and environmental protectors.

The History

The tortuous history began when refiners created cat reformers as a method of raising both the volume and quality of gasoline. In the first half of the 20th century, the demand for gasoline grew at twice the rate of fuel oils. The specifications of the gasoline that car manufacturers designed into their vehicles increased incessantly. America built four and six lane highways across the continent and consumers moved beyond Henry Ford's Model T to big, fat comfortable passenger cars. Their engines needed cheap, higher octane gasoline.

As the refiners scrounged around their refineries looking for suitable blending components, they found heavy naphtha that was purposely left in kerosene. Catalytic reforming could upgrade the quality of these naphthas, some with octane numbers in the range of 35 to 40, to as much as 90 octane, increasing the quality and quantity of the gasoline-making capacity at the same time.

The Golden Years. In 1949, Universal Oil Products Company introduced the present day catalyst and plant design. They arrived just in time for the Golden Years, so-called because of the quarter century following World War II of sustained 7% economic growth in the US Throughout this period, refiners found themselves in an octane race, competing by advertising gasoline with ever-higher *octane numbers.* The gas pump with 100 octane premium gasoline seemed hydrocarbon's Holy Grail. As a way to manage the truth in advertising, more refiners built new and improved cat reformers, boosting the quality of their gasoline even further.

The environment. In the 1970s, enough concern about the environment reached the public agenda that governments began requiring refiners to reduce the amount of lead put in gasoline. For decades, refiners had exploited the mysterious fact that the addition of tiny amounts of tetraethyl lead could substantially increase the gasoline octane. They had always known that lead was toxic, but in the mid-1970s the government published a schedule to phase out in a period of 10 years the already restricted amount of lead allowed. Refiners built new cat reformers, de-bottlenecked old ones, and introduced new catalysts into their existing ones to increase the amount of aromatics, which were high octane gasoline blending components.

Beddour's Law—the premise that one cannot eliminate a pollutant without creating another one[1]—came into effect as lead phased out in the 1980s. The benzene content of gasoline became a big issue. By then benzene was well known as a carcinogen and even the few percent that came primarily from the cat reformers became the focus of concern and eventually phase-down regulation. Once again the refiners went to work to change the feeds and modify the catalysts in the reformer to eliminate this pollutant, replacing it with some other benign, albeit mutated (cat reformed) molecules. Complicating the effort were the limits put on the total aromatics content, since these compounds, with the characteristic benzene ring imbedded in them, were suspect as well.

The by-product workhorse. Finally, throughout the evolution of cat reforming, refiners became increasingly aware that one of the reformer by-products, hydrogen, emerged as an essential workhorse in the refinery. Hydrogen was increasingly used in hydrotreaters to help remove sulfur and other contaminants from various dirty streams and change the structure of some others. Even more, hydrocrackers had become a black hole for hydrogen. Now when the cat reformer shuts down for maintenance or has an emergency shut down, major parts of the refinery get affected by the shortage of hydrogen and have to shut down as well.

The Chemical Reactions

You may want to use that figurative finger stuck in chapter 5, The Chemistry of Petroleum, to move through this chapter without getting too lost.

Unlike the previous processes discussed, boiling range of the feed differs very little from the product of a catalytic reformer. What does change is the chemical composition.

Refiners originally designed the cat reformer to run on straight run naphtha. Naphtha from hydrocracking can also benefit from the changes that take place in the cat reformer. These naphtha streams typically have a high concentration of normal paraffins and naphthenes. The cat reformer causes many of these components to be *reformed* into isoparaffins and aromatics that have much higher octane numbers.

Cat reforming typically results in the material balance in Table 9-1.

	% Volume	
	Feed	Product
Paraffins	45	20
Iso-paraffins	5	15
Olefins	0	0
Naphthenes	40	10
Aromatics	10	55
Hydrogen	0	2

Table 9-1—Reformer Material Balance

The good reactions that take place in the cat reforming process are mainly:

- Paraffins are converted to isoparaffins
- Paraffins are converted to naphthenes, releasing hydrogen
- Naphthenes are converted to aromatics, releasing hydrogen

Some not-so-good reactions, from an octane point of view, take place:

- Some of the paraffins and naphthenes crack and form butanes and lighter gases
- Some of the side chains get broken off the naphthenes and aromatics and form butanes and lighter gases

The important thing for you to remember is *paraffins and naphthenes get converted to aromatics and isomers*, as shown in the reactions in Figure 9-1.

The Hardware

You might expect some unusual hardware would be required to cause these complicated reactions to take place. On the contrary, what's needed is an *unusual catalyst*, and in this case it's made of alumina, silica, platinum, and sometimes palladium. The platinum is in no small amounts (several million dollars worth in one process unit), so great care is taken to keep track of it. The platinum and palladium are the key ingredients that do the wonderful job of causing paraffins to wrap themselves around in circles and to lose their extra hydrogens and to poke side chains out where none existed before.

There are several ways of putting the hydrocarbon in contact with the catalyst. The one covered here is called *fixed bed* because the hydrocarbon is dribbled through the catalyst, which stays put in a vessel, or actually several vessels used in a series, as shown in Figure 9-2.

The operating conditions that promote each of the chemical reactions in Figure 9-1 are different, as measured by pressure, temperature, and residence time. For that reason, cat reformers typically have three reactors, each one doing a different job. The reactors operate at 200-500 pounds per square inch absolute (psia) pressure and 900-975°F. The vessels are characteristically spherical in shape.

The naphtha feed is pressurized, heated, and charged to the first reactor, where it trickles through the catalyst and out the bottom of the reactor. This process repeats in the next two reactors. The product then runs through a cooler where much of it is liquefied. The purpose of the liquefaction at this point is to permit separation of the hydrogen rich gas stream for recycling. This process is important enough to warrant a few sentences.

Look back at the chemical reactions in Figure 9-1. The ones that create cyclics and aromatics result in the production of extra hydrogen because aromatics don't need as many hydrogens as naphthenes and

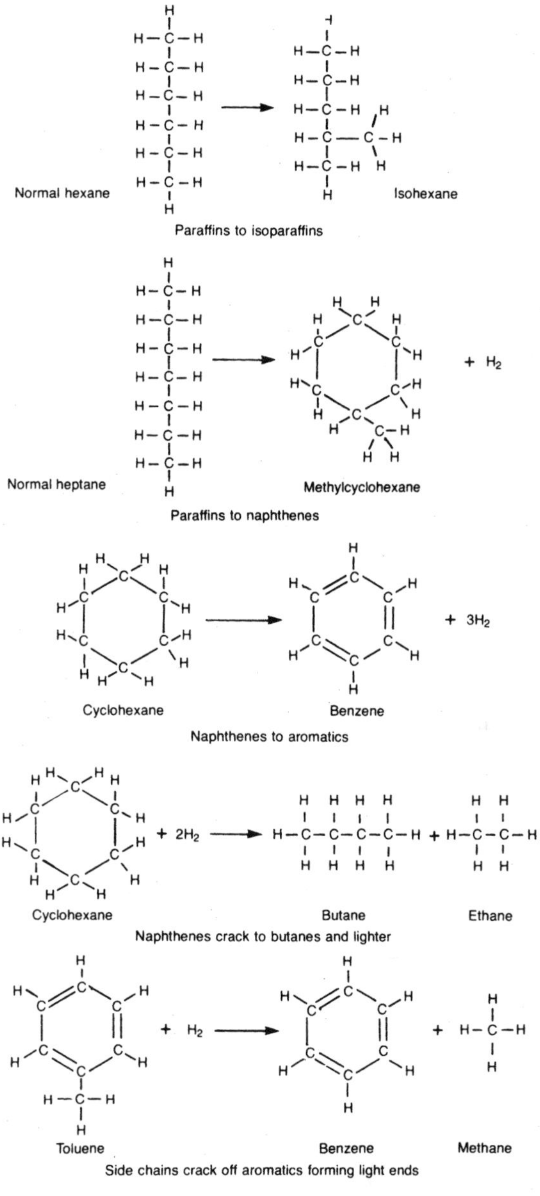

Normal hexane → Isohexane
Paraffins to isoparaffins

Normal heptane → Methylcyclohexane + H_2
Paraffins to naphthenes

Cyclohexane → Benzene + $3H_2$
Naphthenes to aromatics

Cyclohexane + $2H_2$ → Butane + Ethane
Naphthenes crack to butanes and lighter

Toluene + H_2 → Benzene + Methane
Side chains crack off aromatics forming light ends

Fig.9-1—Reformer Reactions

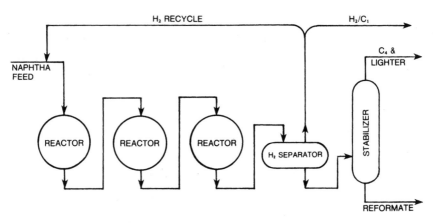

Fig.9-2—**Catalytic Reforming**

paraffins. Reforming is also a user of hydrogen in the reactors. In the chaos that takes place in the reformer, some miscellaneous carbon atoms crack off bigger molecules. Instead of depositing on the catalyst as it does in cat cracking, the carbon reacts with the hydrogen and forms a hydrocarbon gas. Hydrogen is mixed with the feed to keep the concentration of hydrogen vapor in the reactors high enough.

Meanwhile outside the reactors, part of the hydrogen stream is recycled to the feed while the other part is sent to the gas plant. The liquid product from the bottom of the separator is sent to a fractionator called a *stabilizer*, a column nothing more than a debutanizer. It makes a bottom product called *reformate*; butanes and lighter go overhead and head towards the sats gas plant.

Regeneration

As the process proceeds, some carbon in the form of coke does deposit on the catalyst, causing a decline in its performance. Symptoms of the deterioration of *catalyst activity* are either reduced octane numbers, lower reformate yield per barrel of feed, or both, as shown in Table 9-2.

After about 12 months catalyst activity in the early vintage reformers declined so severely that refiners shut them down for 20 or 30 days to regenerate the catalyst. By the 1960s refiners figured out that if they added a fourth reactor, they could take one reactor off line at a time and regenerate it. Since the catalyst in each reactor could be regenerated as frequently as they wanted, the regeneration could be much milder and as short as 30 hours. That kept the reformer going for as much as 36 months without a shutdown. Beside the shutdown costs, that saved refiners a lot of upset in the rest of the refinery, especially from the loss of hydrogen. They could have built stocks of high octane components to last during a reformer shutdown, but hydrogen couldn't be stored in a refinery.

| | Yields-% Volume | |
	Fresh or Regenerated Catalyst	Catalyst after a long run
H_2	2	2
C_1/C_2	2	3
C_3	2	3
iC_4	3	4
nC_4	3	4
Reformate	88	84
Octane number	94	92

Table 9-2—Reformer Yields

In the later vintage reformers, at any one time three reactors are in operation, with the fourth in a regeneration mode. Regeneration is accomplished by blowing hot air into the reactor to remove the carbon from the catalyst by forming carbon monoxide and dioxide. A small amount of chlorine in the hot air will also remove some of the metals deposits. The cycle time for a reactor to be off line is only about 30 hours, thus the catalyst is kept fresh virtually all the time.

Despite the continuous regeneration, over a long period of time the activity of the catalyst will decay. The high temperatures required for regeneration cause the catalyst's pores to collapse; some metals like vanadium or nickel deposit on the catalyst. Consequently, every 2-3 years the entire reformer must be shut down to swap out the catalyst for some fresh stuff. The platinum and palladium in the catalyst are the expensive parts. The refiners either own or lease these precious metals; the processors just do the refurbishment for a fee, returning the same amount of metal to the refiners.

Process Variables

The operators play with a variety of dials and buttons when they run the reformer—temperature, pressure, residence time, feed quality, feed cut points, and more. The game is really a balance between the volume of

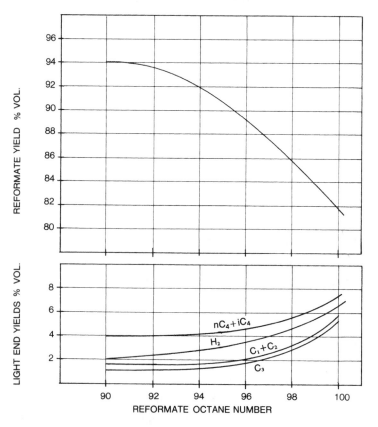

Fig.9-3—**Cat Reformer Yields vs. Octane Number**

the reformate produced and its quality. Figure 9-3 shows one of the relationships—*as the octane number goes up, the percent of volume reformate goes down.* Correspondingly, the yield of butanes and lighter goes up. That happens because more reforming of the heavier molecules inevitably leads to side chains breaking off or free carbon atoms forming methane or ethane. The operation of the cat reformer must be tuned very closely to the gasoline blending operations and the gasoline component yields of the other processing units.

The properties of the naphtha feed, as measured by the *paraffin, olefin, naphthene,* and *aromatic* (PONA) content, effect the yields and quality also. The PONA analysis of naphtha is an important component in the analysis of a crude oil's value. If you hold a lot of other things constant, the more naphthenes in the feed, the more aromatics are likely to be in the product; the more normal paraffins, the more isoparaffins come out.

Cat reforming is a primary source for benzene, toluene, and xylene. You can read about the process for recovery of the BTXs in chapter 18.

Review

Catalytic reforming is an important process for upgrading low octane naphtha to a high octane gasoline blending component, reformate. The paraffins are converted to isoparaffins and naphthenes are converted to aromatics to capitalize on their higher octane numbers. Unfortunately, the higher the octane number of the reformate, the lower the yield and the more light ends produced.

Exercises

1. Fill in the blanks:
   a. The basic purpose of catalytic reforming is to increase the _____ of naphtha.
   b. The expensive components of the reforming catalyst are _____ .
   c. Reformate has a much higher concentration of _____ and _____ than the reformer feed.

    d.   An important by-product of cat reforming is _____.

    e.   The other by-products of cat reforming are _____, _____, _____, and _____.

    f.   To keep the catalyst from getting coked up, the reactors are kept filled with _____.

    g.   Symptoms of aging catalyst are the decline of the reformate _____ or _____, and the increase yield of the _____.

2.   Suppose the reformer feed is 15 MB/D and the operating conditions are set to produce 91 octane reformate. The 91 octane reformate is worth 100 cents/gal and the C_4 and lighter streams are worth 50 cents/gal. Each increase in octane number increases the reformate value by 1 cent/gal. Does it make sense to increase the severity of the reformer, that is, run it to produce higher octane reformate? Use the yield versus octane number chart in Figure 9-3. (Hint: check the economics at 91, 95, and 100 octane).

3.   Draw an additional example of each of the five types of reactions that take place in the catalytic reformer.

4.   In the 1960s, Shell advertised the use of Platformate® in its premium gasoline. What is Platformate® a contraction of?

5.   Draw the refinery configuration covered so far, including the cat reformer in its correct place.

References

[1] Dr. Robert F. Beddour, Professor of Chemistry at the University of Texas.

10
Residue
Reduction

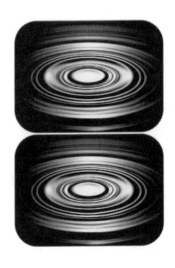

The very ruins have been destroyed.

-Civil War, Luca

H igh rates of crude runs inevitably produce large volumes of the *bottom of the barrel*. During the 20th century, the growth in gasoline and the lighter end of the barrel has outstripped the growth in heavy fuel oil. The developed countries led; the developing countries followed. Even the reduction of straight run residue volumes by vacuum flashing could not balance residual fuel supplies with demands. So refiners have resorted to various technologies to convert residue to light products. As early as 1920, large volumes of pitch were being processed in Dubbs Units, which were *thermal crackers* named after their designer. These produced distillate fuels and low quality gasoline, but they greatly improved the gasoline/residual fuel imbalances. After vacuum flashing and cat cracking appeared on the scene, refiners used another form of residue reduction called visbreaking to "loosen up" the flasher bottoms so they could sell them as fuel oil. Cokers, a later innovation, converted residue to light products and coke. More recently catalysts have been developed to permit residue cat cracking, a similar but not nearly as effective process for producing light products as its predecessor, conventional cat cracking. But it does reduce the volume of residue coming out of the refinery. These processes are very similar to each other, which makes it apropos to cover them all in one chapter.

It's important to note that there is sometimes a fuzzy understanding of the difference between *residue reduction* and *pitch destruction*. The former does not always imply the latter. There will be more on this subject at the end of the chapter.

Thermal Cracking

In the chapter on distilling you discovered that heating petroleum to temperatures in the 1000°F range causes molecules to break up. In the chapter on vacuum flashing, you learned about a process unit designed to *avoid cracking* of molecules due to high temperature. In cat cracking you found a way to control cracking using high temperatures and catalysts. In this section on thermal cracking, you will learn about *promoting cracking* by high temperatures without catalysts.

The process. The furnaces heat the feed to the 950-1020°F range (Fig. 10-1). Residence time in the furnaces is kept short to prevent much reaction from taking place in the tubes going through the furnace. Otherwise coke will form, quickly clogging the furnace tubes, shutting the operation down. The heated feed is then charged to a *reaction chamber*, which is kept at a pressure high enough (about 140 psia) to permit cracking but not much coking.

The product from the reaction chamber is *quenched* to stop the cracking by mixing it with a somewhat cooler recycle stream. The effect is similar to throwing a pail of water on two amorous dogs.

Both streams are then charged to a *flasher chamber*, where the lighter products go overhead because the pressure is reduced as in a vacuum flasher. The lighter products from the top of the flash chamber are charged to the fractionator, as shown in Figure 10-2. The C_4 and lighter streams are sent to the cracked gas plant. The thermal cracked gasoline and naphtha are used for gasoline blending. The gas oils can be used as

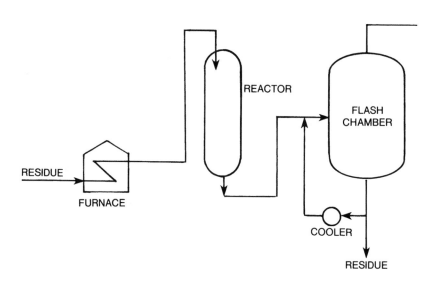

Fig. 10-1—Thermal Cracker Reaction Section

a distillate fuel. The fractionator bottoms are a heavy residue, part of which is the quench stream or recycled to the furnaces; the balance is usually blended into residual fuel.

Product quality. The thermal cracker generates products that have different properties than distilling, more like cat cracking. The feeds, typically flasher bottoms or maybe some cycle oil from the cat cracker, are loaded with big, complex molecules. They have a heavy dose of multiple aromatic ring compounds and very low hydrogen-to-carbon ratios. While long chain paraffins can rupture anywhere, the cyclic compounds tend to break at the point where a straight chain (if any) is attached. With the shortage of hydrogen, this all means that thermal cracking tends to produces molecules long on double bonds, either olefins or compounds with aromatic rings.

The boiling ranges of the thermal cracked gasoline and the thermal cracked light gas oil are suitable for use in the gasoline blending pool and the distillate fuel blending pool. However their olefinic and aromatic con-

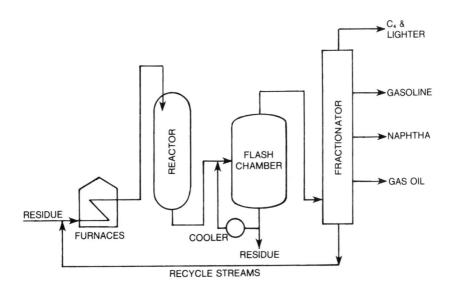

Fig. 10-2—Thermal Cracker

tents may have to be handled carefully. You can read more about that in chapters 12 and 13. Sometimes the thermal cracked naphtha will be upgraded in a hydrotreater or in the cat reformer—the thermal cracked light gas oil can be treated in a hydrotreater or a hydrocracker, if there is one.

Coking

If thermal cracking is like throwing a hamburger on a hot grill, coking is like Texas barbecue—slow cooked all the way through. Through the years, as refiners gathered more knowledge about thermal cracking, they found they could raise heavy feedstocks to high temperatures without coking up their furnace tubes if they used high velocities to postpone the formation of coke. They then could let the coking take place in a vessel where they could remove it relatively easily. The furnace tubes through which the feed passes to be heated are fairly narrow to allow efficient heat exchange. Coking in these tubes is always a disaster for refinery maintenance people. Better the coke formation take place where it can be more easily dealt with.

Using a wider diameter vessel called a *coke drum* made the removal of the coke feasible. Still, the hardware is mechanically more complicated than refiners like, since they are, for the most part, fluids handlers. Coke is a solid and the *coke handling* facilities presented some unique challenges. Eventually they developed designs to handle the whole process efficiently on a continuous flow basis.

Coker design has gone down three paths—*delayed cokers, fluid cokers,* and *flexi-cokers.* Not many of the last two have been built, so this chapter will concentrate on the first, delayed cokers, and just describe the main differences with the other two at the end of the section.

The process. The feed to the delayed coker (same ones as a thermal cracker-flasher bottoms, cycle oil) is quickly heated to about 1,000°F and then charged to the bottom of a coke drum (Fig. 10-3). The cracked lighter product in vapor form rises to the top of the drum and is drawn off and sent to a fractionator for separation, just like the thermal cracker.

The heavier product remains in the bottom of the insulated coke drum and continues to crack until all the molecules containing hydrogen crack off. The carbon left behind, called coke, is in solid form.

Coke removal. Back in the old days, the thermal cracker reaction chamber sometimes got *coked up* because of some upset or accident. The only way to get the coke out was to send workers into the vessel with chipping hammers and oxygen masks. Surely this is what inhibited the development of coke production in refineries.

Currently, *decoking* is a routine daily occurrence and is usually accomplished by using a high-pressure water jet (about 2,000 psia). First a hole is drilled in the coke from the top to the bottom of the drum. A rotating stem is then lowered through the hole, spraying a water jet sideways. The high pressure cuts the coke into lumps, which drop out the bottom of the drum into trucks or rail cars for shipment or hauling to the *coke barn.*

Coke drums typically come in pairs and run on a 48 hour cycle. Filling a drum with coke takes about 24 hours. Switching, cooling, decoking, and emptying take about 22 hours, during which time the other

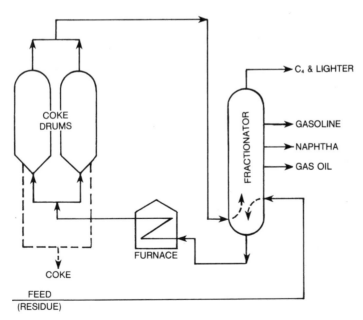

Fig. 10-3—Coker

drum is filling. Alternating drums makes delayed coking seem like a continuous process.

Yields. The outturns from thermal cracking and coking are sensitive to operating conditions, mainly furnace temperature, and properties of the feed. Boiling range, PONA analysis, and gravity are usually used to predict yields. As an example of what these units can do, feeding flasher bottoms from West Texas crude to a thermal cracker being run to maximize gasoline or to a coker might operate as follows:

	Thermal Cracker Yields	Coker Yields
Pitch Feed	1.00	1.00
Products		
Coke*		.30
C_2 and lighter	.18	.15
Propane	.05	.01
Propylene	.07	.01
Butanes	.04	.02
Butylenes	.06	.04
Gasoline	0.30	0.08
Naphtha	0.15	0.15
Gas Oil	0.10	0.50
Residue	0.20	—

% weight, since it's not liquid.

The thermal cracker makes about 80% of the residue go away. The coker makes it all go away, but it also makes 30% coke. Both processes also make gasoline, naphthas, and gas oils that have properties, advantages and disadvantages similar to cat cracked streams. Coker naphtha can be upgraded in a hydrotreater or the cat reformer; coker gas oil is sometimes further processed in a hydrotreater or a hydrocracker, if there is one. However, residue goes away, and that's the name of the game!

Coke. The unique characteristics of coke warrant a few words. Most coke produced in refineries is sponge-like in appearance, if you look closely; hence the name *sponge coke*, or sometimes *green coke*. Most sponge coke ends up as fuel in cement kilns or in power plants where it is mixed with coal and burned.

Coking the residue of some types of crudes produces coke pure enough for a variety of industrial products. Manufacturers of silicon carbide and calcium carbide use coke as a raw material. The aluminum industry uses *anode grade coke* to make carbon anodes that convert alumina to aluminum. In order to make it suitable for these uses, the coker coke is cooked in a *calciner*. Refiners use this alliterative process because the coke, as it leaves the coker, still contains some volatiles and even water trapped in the solids. *Calcinable coke* needs to be very low on contaminates, especially sulfur and heavy metals such as vanadium and nickel which are common in many crudes.

Only a few crudes can be used to make an exotic-sounding variety of calcinable coke called *needle coke*. The name comes from the elongated crystalline structure. Needle coke is used to make anodes for electric arc furnaces, mostly for steel making. The special structure of needle coke lets it handle temperatures up to 5,000°F without falling apart. Since the crudes are few and the process is touchy, needle coke commands a higher price premium than regular calcinable coke, which itself sells for a higher premium than sponge coke.

Different refiners run their cokers to make money in their own ways. Most run them to maximize the production of light oils and end with sponge coke, the cheapest of the varieties. Some run them to make the more valuable calcinable coke, and so they take care to select the right crudes and run the operation at higher pressures. Only a few refiners select crudes and operate their cokers to produce needle coke. Even then, the refiners can be so selective as to produce *intermediate*, *premium*, or *super premium grade* needle coke, all commanding the elevated prices their names imply.

Visbreaking

A visbreaker is a poor man's thermal cracker. On paper it looks like a thermal cracker, but the hardware is simpler and cheaper to build and operate. On the other hand, only about 20-30% of the pitch is destroyed.

Usually the *pitch* made in the refinery (the flasher bottoms or the cat cracked cycle oil) is not marketable by itself because it is so viscous. *Cutter stock*, such as cat cracked heavy gas oil or even some of the heavy flashed distillate, can be added to make it flow at reasonable temperatures. Unfortunately, adding cutter stock makes more volume of a product that commands about the lowest price of all refined products. (There's more on viscosity and residual fuel in chapter 14).

A visbreaker takes pitch and does a mild thermal cracking. The lightest of the cracked product is allowed to flash off in a flash chamber, similar to that in Figure 10-1. The rest, left in the pitch as a *diluent*, eliminates the need to add cutter stock. Some visbreakers have no flash chamber. The pressure release in the valve between the reactor and the fractionator lets the lighter hydrocarbon flash sufficiently as it enters the fractionator.

You might think of visbreaking as adding Ex-Lax® to the refinery, loosening up the bottom of the barrel. The only difference is a net *reduction* in residue produced. The refiner doesn't have to build either an expensive coker or a full-scale thermal cracker to pull it off.

Other Residue Reduction Options

Fluid coking. In a process that looks like cat cracking on paper, the fluid coker makes coke "on the fly" and separates the coke from the hydrocarbon using cyclones, *a la* the cat cracker. The process is difficult to control and most refiners have chosen to go the delayed coker route. The fluidized coke is used as a fuel for steam generators that have to be located nearby to keep the coke moving.

Flexi-coking. Another design that has not yet captured the imagination of most refiners is flexi-coking. The process ends with almost all the coke converted to CO. The CO is mixed with the C_2 and lighter by-products to produce a low quality fuel gas that can be marketed to power plants.

Residue cat cracking. A few refiners have built cat crackers designed to use the residue as feed. Relative to a conventional cat cracker feeding straight run heavy gas oil and flashed distillate, everything is larger in a residue cat cracker. The risers have to be wider and longer to allow more residence time, the regenerators are larger because the relative amount of coke laid down on the catalyst is higher. The catalysts are designed to handle more complex molecules in their pores, and the crudes are chosen to assure they do not have contaminants that poison the catalyst.

Comment. How does a refiner choose which residue reduction process to put in a refinery? The choices depend on many factors, including the value of residual fuel, the value of the light products, the value of coke, the capital and operating costs of the various hardware configurations, and the properties of the crude oil being run in the refinery. High metals content, high carbon-to-hydrogen ratio, and high coke prices push a refiner in the direction of delayed coking and away from residue cat cracking. Poorer coke prices but better residual fuel prices push toward thermal cracking and visbreaking

Exercises

1. Fill in the blanks.

   a. The main difference between thermal cracking and cat cracking is that the former uses no _____ to promote cracking.

   b. The main difference between thermal cracking and coking is the production of _____ and _____.

   c. To prevent coking in a thermal cracker reaction chamber, the feed is _____.

   d. To promote coking in a coke drum, the feed is _____.

   e. Thermal cracker and coker gasoline, naphthas, and gas oil are _____ in quality.

   f. The various names refiners give to different types of coke are _____, _____, _____, _____, and _____.

   g. The thermal cracked and coker C_4 and lighter gas is sent to the _____ gas plant because it contains _____.

    h.   The three main parts of thermal crackers or cokers are
the _____, _____, and the _____,
which in a coker is called a _____ _____.

2.   The refinery manager in chapters 2 and 3 has to tell the sales manager how much coke to sell. He runs all the flasher bottoms to the coker. Use the yields in this chapter and answers to problem 2 in chapter 3 to do the calculation. Assume the gravity of the flasher bottoms is 10°API or 350 lb/bbl. (Remember—the coke yield is percent weight, and the units of sale for coke are short tons, 2,000 lb per ton).

3.   Add a thermal cracker to your drawing of the refinery process units and stream flows.

11

Hydrocracking

Cancel and tear to pieces that great bond.

-Macbeth, Shakespeare

Hydrocracking is a process of more recent vintage than thermal or cat cracking or cat reforming, but it was designed to accomplish more of what each of the earlier processes do. Plopped in the middle of a refinery, *hydrocracking* can take care of many refiners' headaches that happen as the market changes from month-to-month or season-to-season. Hydrocrackers can produce gasoline components from light or heavy gas oils. Their quality is better than if those gas oils were recycled through the cracking process that generated them. Hydrocrackers can produce light distillates (jet fuel and diesel fuel) from heavy gas oils. Hydrocrackers produce a relatively large amount of isobutane, a handy supply for the alky plant. Maybe best of all, hydrocrackers produce no bottom-of-the barrel leftovers (coke, pitch, or resid). The outturn is all light oils. Refiners use hydrocrackers to move from max diesel and distillate fuels in the winter to max gasoline and maybe even jet fuel in the summer. In the refineries where they reside, hydrocrackers have become refiners' swing units.

The Process

Hydrocracking is simple. It's catalytic cracking in the presence of hydrogen. Various combinations of hydrogen, catalyst, and operating conditions permit cracking a wide range of feedstocks, from light gas oil to straight run residue or the cycle oils from the cat cracker or thermal cracker. The hydrocracker is run in stages, with each one upgrading to the next cut—heavy stuff to middle distillates, middle distillates to gasoline range components.

Hydrocracking can simultaneously improve the quality of both the gasoline blending and the distillate fuel blending pools. The worst of the distillate stocks, the cracked gas oils, have high aromatics contents that give them poor diesel fuel performance. Passing them through a hydrocracker results in gasoline components with relatively high octane numbers or naphthas that make excellent cat reformer feed.

Why doesn't every refinery have one of these machines? Even though about a dozen different types of hydrocrackers are presently popular, they all are expensive to build and operate. The units described below are typical of most of them.

The Hardware and the Reactions

Hydrocracking catalysts are fortunately less precious and expensive than reforming catalysts. Usually they are compounds of sulfur with cobalt, molybdenum, or nickel plus alumina. (You may have wondered what anyone used those metals for.) In contrast to cat cracking, but like cat reforming, hydrocrackers have their catalysts in a fixed bed. Like cat reforming, the process is carried out in more than one reactor—two in the illustration in Figure 11-1.

Feed, in this case cracked heavy gas oil, is mixed with hydrogen vapor, heated to 550-750°F, pressurized to 1,200-2,000 psi, and charged to the first stage reactor. As it passes through the catalyst, about 40-50% of the feed is cracked to gasoline range material (below 400°F end point).

The hydrogen and the catalyst are complementary in several ways. First, the catalyst causes long chain molecules to crack and the rings in aromatic compounds to open. Both these reactions need heat to keep them going. They are both an *endothermic process.* On the other hand, as the cracking takes place, the excess hydrogen floating around saturates (fills out) the molecules, a process that gives off heat. This process, called *hydrogenation,* is *exothermic.* Thus, hydrogenation gives off the heat necessary to keep the cracking and ring opening going.

The catalyst and hydrogen also complement each other in the formation of isoparaffins. Cracking forms olefins, which could join together to form normal paraffins. Hydrogenation rapidly fills out all the double bonds, often forming isoparaffins, preventing reversion to less desirable molecules. (Isoparaffins have higher octane numbers than normal paraffins.)

After the hydrocarbon leaves the first stage, it is cooled so most of the hydrocarbon liquefies. In the *hydrogen separator* the hydrogen is split out for recycle to the feed. The liquid is charged to a fractionator. Whatever products are desired (gasoline components, jet fuel, and gas

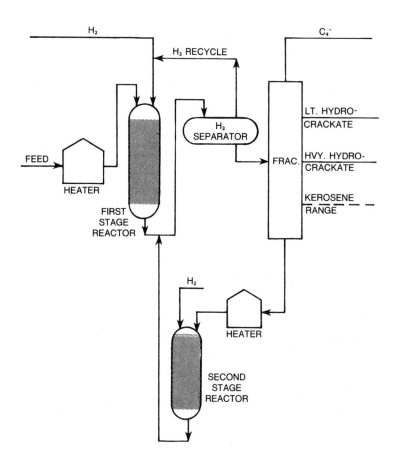

Fig. 11-1—**Two Stage Hydrocracker**

oil), the fractionator cuts them out of the first stage reactor effluent, leaving a bottoms stream ready for the next, second stage, reactor. In other words, kerosene range and light gas oil range material could be taken as separate side draw products from the fractionator, or could be included in the fractionator bottoms if further conversion to gasoline range material is the object.

The bottoms stream is again mixed with a hydrogen stream and charged to the second stage. Since this material has already been subjected to some hydrogenation, cracking, and reforming in the first stage, the operations of the second stage are more severe (higher temperatures and pressures). Like the outturn of the first stage, the second stage product is separated from the hydrogen and charged to the fractionator.

Some hydrocrackers are configured in three stages, either by stacking different catalysts in one reactor or by having three reactors. Different reactions take place in each stage. In the first stage, the catalyst opens up the rings of complicated molecules where the contaminates, sulfur and nitrogen, might be embedded. The hydrogen forms hydrogen sulfide (H_2S) and ammonia (NH_4). At the same time the hydrogen fills out many of the opened double bonds, forming simpler, lighter compounds.

In the second stage, run more severely, the catalyst opens more rings of the heavy, complicated molecules and cracks others, forming light products. In the third stage, a kind of polishing job takes place where the olefins and aromatic compounds are saturated, forming naphthenes, paraffins, and especially isoparaffins.

Hydrocracker reactors operate at really severe conditions—2,000 psi and 750°F. Imagine the hardware necessary to contain the reactions. The specialty-steel reactor walls are sometimes 6 inches thick. A critical worry is the possibility of *runaway cracking*. Since the overall process is endothermic, the temperature can rise rapidly, accelerating the cracking rates dangerously, which only gives off more heat as hydrogenation then takes place. Elaborate quench systems are built into most hydrocrackers to control runaway.

Yields. Another excellent feature of the hydrocracker is a 10-20% gain in volume. The cracking/hydrogenation combination results in products whose average gravity is a lot higher than the feed. The table below shows typical yields from hydrocracking cat cracked cycle oil and cat cracked light gas oil. The primary products, blending components heading for the gasoline pool, are called *hydrocrackate*.

	% Volume Balance	
Feed		
Cat cracked cycle oil	1.00	
Cat cracked light gas oil		1.00
Product		
Propane	0.02	0.03
Iso-butane	0.02	0.02
Normal butane	0.06	0.08
Hydrocrackate	0.70	0.80
Kerosene range	0.10	0.17
Light gas oil range	0.15	0.00
Total	1.15	1.10

Table 11-1—**Hydrocracker Yields**

Not shown is the hydrogen requirement, which is measured in standard cubic feet (scf) per barrel of feed. Net consumption of 2,500 scf/bbl is typical.

The heavier part of the hydrocrackate contains a lot of aromatics *precursors* (compounds that easily convert to aromatics in a reformer). Refiners often separate this naphtha from the light hydrocrackate and feed it to the reformer for upgrading. The kerosene range material makes a good jet fuel or a distillate blending stock because of its low aromatic content, a consequence of hydrogen saturating those double bonds. Chapter 13, on distillate fuels and chapter 14, on hydrotreating, will explain why this is good.

Residue hydrocracking. A handful of hydrocrackers have been designed and constructed to use straight run residue or flasher bottoms as feed. Most of them are operated as hydrotreaters, as described in chapter 14. The yields are more than 90% residual fuel. The purpose of the operation is to remove sulfur by the catalytic reaction of the hydrogen and the sulfur compounds, forming H_2S. Residue with a sulfur content of about 4% or less can be converted to heavy fuel oil with less than 0.3% sulfur.

Only a few real residue hydrocrackers producing predominantly light products have been built. They pump enormous amounts of hydrogen into the reactions to break up the very heavy molecules, and require lots of recycling of the heavy ends back through the reactor to make them go away.

A residue hydrocracker costs a relatively large bundle of money just to convert high sulfur residual fuel to low sulfur residual fuel, and even more to convert it to light products. Special circumstances including location, access to cheap, heavy crude oil, and lack of any high sulfur fuel market might do it.

Review

With the addition of the hydrocracker to the refinery processing scheme, the absolute requirement for integrated operations becomes apparent. The hydrocracker is a pivot point since it can swing refinery yields between gasoline, distillate fuel, and jet fuel and simultaneously improve the quality. How the hydrocracker runs depends intimately on the feed rates and operating conditions of the cat cracker and residue reduction units—the coker or thermal cracker. In addition, the hydrocracker feeds the alky plant with iso-butane and the cat reformer with naphtha and gets hydrogen back from the cat reformer.

Exercises

1. Contrast hydrocracking, cat cracking, and thermal cracking in terms of feed, what promotes the reaction, and what the products and product PONAs are.

2. How is a hydrocracker complementary to a cat cracker? A reformer to a hydrocracker?

3. Draw the refinery flow diagram with the hydrocracker added.

chapter twelve

12

Gasoline

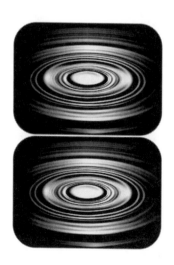

The good things in life are not to be had slightly, but come to us with a mixture.

-Charles Lamb

F inally you get to gasoline, the subject everybody knows a little about. Gasoline is the product almost everybody buys and occasionally spills on their shoes. They know what yesterday's gasoline price was because it's posted on every other corner. They all understand that higher octane is better than lower octane, but they probably don't know why. That's all okay for most people because a competitive marketplace delivers gasoline grades that will work just fine in any car. Most people don't need to understand much more—except you.

It used to be, in earlier editions of this book, that understanding how to make gasoline was a fairly simple matter. However, since the 1980s, governments and institutions have learned how bad "old fashioned" gasoline can be for your health. Governments, especially in America, have successively put more restrictions on the content of gasoline, giving progressively bigger headaches to refiners. The best way to cover the accumulated complexities is with a historical tour, starting with what refiners think of as "the good old days" and environmentalists consider the "black past." So this chapter will start with gasoline in the 1960s and work its way to the 21st century, covering these questions:

- What happens when gasoline is burned in a car engine?
- What is octane, and the equally important property, vapor pressure?
- Whatever happened to tetraethyl lead, and what did that do?
- How does gasoline effect the environment?
- How do refiners blend gasoline now to not affect the environment?
- What effect does gasoline blending have on the way to run a refinery?

Gasoline Engines

The essential parts of a gasoline engine, at least for this discussion, are the gas tank, the fuel pump, fuel injection, the cylinders, the pistons, and the spark plugs. Engines without the last item (spark plugs) will be discussed in the next chapter because they're diesels.

You might say the whole motive process in a car starts at the gas tank when you fill it at a gas station. As you start the engine, the fuel pump sucks gasoline out of the tank and sends it to the fuel injection apparatus.

The purpose of fuel injection is to get the gasoline in the right state and in the right place to burn it, releasing the chemical energy. Gasoline, like other hydrocarbons, doesn't burn in its liquid state. It has to be vaporized and mixed with oxygen to ignite. For example, when you throw charcoal lighter on the coals in your barbecue grill and toss in a match, the vapors ignite, not the liquid. That's why some dummies can get away with squirting charcoal lighter directly on a lit fire without barbecuing themselves. The liquid stream doesn't ignite. Unless the charcoal lighter has been sitting in the sun or next to the grill and is warm, the stream doesn't vaporize in the air, only when it hits the hot coals.

The fuel injection apparatus mixes the gasoline with air, pumps it up to a high pressure, then squirts it into the cylinder. The sudden release of pressure is enough to vaporize the mixture, filling the cylinder with a gasoline/air mixture, ripe for ignition. The nitrogen that's also in the air just passes through the process, more or less unaffected. (Less is bad, as you'll read in a few pages).

The sequence of happenings illustrated in Figure 12-1 shows one full cycle of a four cycle engine. Keep in mind you may have four, six, or eight cylinders doing this in your car, cycling at 2-3,000 times per minute.

The downward movement of the piston works together with the fuel injection apparatus to suck the gasoline/air mixture into the cylinder. At the bottom of the stroke, the space in the cylinder reaches its maximum and is filled with the fuel. The fuel injection apparatus closes and in the next step, *compression*, the piston moves up the cylinder, compressing the vapor. When the piston reaches the top of the stroke at the point of *ignition*, the spark plug gives off a powerful spark, igniting the gasoline vapor. The gasoline burns rapidly, just short of an explosion. The gases expand and put huge pressure on the piston, forcing it down the cylinder in the *power* stroke. If the gasoline is formulated correctly, different molecules will burn at different times so that the combustion takes place over the whole length of the *power* stroke, smoothing out the motion. Power is transmitted to the crankshaft as the piston is forced down the cylinder in the power stroke. At the *bottom* of the power stroke, the exhaust valve at

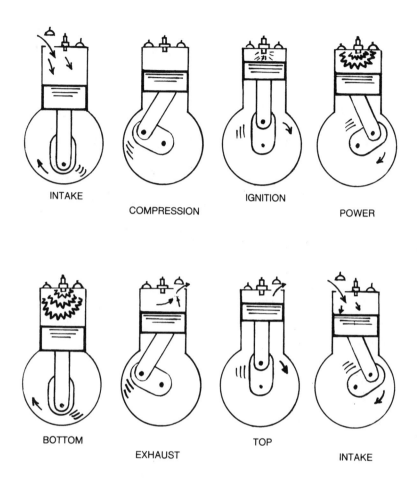

INTAKE

COMPRESSION

IGNITION

POWER

BOTTOM

EXHAUST

TOP

INTAKE

Fig. 12-1—**Four Cycle Internal Combustion Engine**

the top of the cylinder opens. Burnt fuel is pushed out in the *exhaust* stroke as the piston moves up the cylinder. At the *top* of the stroke, the outlet valve closes, the injector opens up again to squirt the fuel in, and the process is ready to be repeated.

Note that each cycle requires two trips of the piston up and down the cylinder, which is not the reason why it's called a *four cycle engine*. The four cycles are intake, compression, power, and exhaust, which aren't really cycles, but nobody ever said automotive engineers were all that articulate.

Some small two cycle engines (lawnmowers, outboards, etc.) use gasoline. The big two cycle engines run on diesel fuel or the heavier residual fuel. You can read about them in chapters 13 and 14.

Vapor Pressure

One of the crucial steps in the engine cycle is the vaporization of the gasoline. When the engine is warm, there is no problem—the engine heat assures that 100% of the gasoline will enter the cylinder in vapor form. When the engine is being started from cold, the injector may spit some droplets instead of vapor into the cylinder, making ignition difficult at worst, less smooth at best.

The trick in handling cold starts is to have enough volatile hydrocarbon in the gasoline to get a vapor-air mixture that will ignite. The measure of volatility is vapor pressure, and more specifically *Reid vapor pressure* (RVP), immortalizing the man who designed the test apparatus.

Definition. Vapor pressure is a measure of the surface pressure it takes to keep a liquid from vaporizing. A light hydrocarbon like propane will have a very high vapor pressure since it is very volatile. A heavier hydrocarbon like gas oil will have nearly zero vapor pressure, since it will vaporize very slowly—at normal temperatures. If you think for a moment you'll realize that any vapor pressure is a function of temperature. RVP is measured at 100°F, an arbitrary number Reid chose.

Engine conditions. Definitions out of the way, go back to the vaporization problem. The RVP of gasoline must meet two extreme conditions. On cold starts, enough gasoline must vaporize to provide an ignitable mixture. Once ignition occurs, the rest of the gasoline will vaporize and burn too. The other extreme happens when the engine is running while completely warmed or even more extreme, is being restarted when it's hot. At that point, the gasoline vapor must not expand in the injection apparatus so much that no air can be mixed with it. Again, the mixture must be ignitable.

Refiners have found that there is a direct correlation between a gasoline's ability to meet those conditions and the RVP. Furthermore, they have found that the ideal RVP for gasoline varies with the seasons. In the dead of winter in a place like Bemidji, Minnesota, cold starts need a gasoline with a 12 psi RVP. During the dog days of August in Brownsville, Texas, cars won't restart if the gasoline has a higher RVP than 8.5 psi.

Vapor lock. One other constraint on vapor pressure is worth mentioning—vapor lock. Car engines can shut down if they encounter an unexpected high altitude or high temperature. At high altitudes, the atmospheric pressure is lower, and high RVP gasoline will tend to vaporize anywhere in the system. High temperatures aggravate the problem. The fuel pump tries to pump a combination of vapor and liquid when it is designed to handle only liquid. Consequently, the whole system will starve, and the engine will quit and won't start again until the temperature of the gasoline is lowered. That could take hours. To avoid vapor lock, gasoline RVP is localized to accommodate the temperature and pressure conditions in the area, including seasonal temperature and barometric swings.

Blending gasoline to meet RVP requirements. So much for what cars need. How do refiners get there? If you look at the list of gasoline blending components in Table 12-1, you'll see that all but three have RVPs below the limits mentioned above. The answer might jump right out at you—as the industry evolved, butane became the pressuring agent of choice.

Blending Components	RVP-psi
iC4	71.0
nC4	52.0
Reformate	2.8
Hydrocrackate	2.5
Alkylate	4.6
Straight-Run Gasoline	11.1
Straight-Run Naphtha	1.0
Cat Cracked Gasoline	4.4
Coker Gasoline	4.0

Table 12-1—Vapor Pressures

If you sat down to design the scheme for the refining industry to blend gasoline, you'd probably reject the thought that there would be enough butane around to be the marginal component used for all vapor pressure control. But, for many years and places that's the way it has turned out. Butanes are made in refineries as a by-product of conversion processes. They also come into the refinery mixed with some crudes and are recovered from natural gas in gas processing plants. Somehow these three rather inelastic supplies end up providing all the butane that's needed for gasoline blending.

Getting down to nuts and bolts, the procedure for calculating the amount of butane needed to pressure up the gasoline involves only algebra and weighted averages. Vapor pressure calculations are not exactly related to the volumetric weighted averages, but for the purposes of this drill, they're close enough. Suppose the RVP specification is 10 psi and you have a blend of five components. Table 12-2 shows how much normal butane is needed.

Component	Volume in Barrels	RVP	Volume x RVP
Straight run naphtha	4,000	1.0	4,000
Reformate	6,000	2.8	16,800
Hydrocrackate	1,000	2.5	2,500
Cat cracked gasoline	8,000	4.4	35,200
	19,000		58,500
Normal butane	x	52	52x

Let x = the amount of butane you need. For a spec of 10 psi RVP

$$10 \times (19,000 + x) = 58,500 + 52x$$
$$190,000 + 10x = 58,500 + 52x$$
$$x = 3,131 \text{ barrels of normal butane required}$$

Total gasoline production is 19,000 + 3,131 = 22,131 bbl.

Table 12-2—Blending for Vapor Pressures

So the calculation is pretty simple, but some of the implications might be mentioned. Since the specification for RVP is higher in the winter than in the summer, the capacity to produce gasoline is higher in the winter. The higher the RVP spec the more butane refiners can blend in and the more total volume of gasoline they make. Unfortunately, in most markets, except those like Miami Beach, Florida or Vail, Colorado, the demand for gasoline is lower in the winter than the summer. The additional winter gasoline blending capacity does, however, give some flexibility to increase distillate fuels production without having to under-manufacture gasoline.

Normal versus isobutane. Why use normal butane instead of isobutane to pressure gasoline? There are several good reasons. First of all, the RVP of normal is 19 psi less than iso (71 versus 52), which means that more normal butane can be blended in. Second, normal butane is more "normal" and plentiful in nature. Third, isobutane has another welcome home, alkylation. Very often refineries don't even have enough isobutane to satisfy the appetite of the alky plant, and they have to feed some normal butane to a *butane isomerization plant* (see chapter 16) to make isobutane. All this results in normal butane prices normally lower than isobutane prices and that gives an incentive to blend as much normal butane as allowed.

Do you recall back in the "good old days" watching a car gas tank as you filled it and seeing a wavy-looking vapor appear around the gas cap. That was butane escaping from the gasoline blend. If you recall correctly, you probably saw more vapors in the winter than in the summer. That's the result of the higher vapor pressure spec and more butane in the winter versus the summer. Since those times, the escaping butane and other hydrocarbons have been identified as major contributors to air pollution and lower RVPs have been mandated to restrict those emissions. Many places have clumsy vapor recovery systems on the nozzle to capture and recycle the vapors.

Octane Number

Everyone knows that bigger and more powerful cars need higher octane, and they accept that higher octane gasoline should cost more.

This section will give you a clue why that is, in fact, true.

Octane numbers are measures of whether a gasoline will *knock* in an engine. That's a fine definition, but it requires an explanation of another almost universally obscure phenomenon—knocking.

Knocking. You'll find it helpful to refer back to Figure 12-1, the diagram of the engine cycle. After the gasoline/air vapor is injected into the cylinder, the piston moves up to compress it. As the vapor is compressed, it heats up. (Ever feel the bottom of a bicycle pump after you've pumped up a tire? It's hot. Same effect as in an engine cylinder.) If the gasoline/air vapor is compressed enough, it will get hot and start to *self-ignite*, without the aid of a spark plug. If this happens before the piston reaches the top of the stroke, the engine will knock and push the piston in the wrong direction, against the crankshaft instead of with it. Typically, knocking occurs only momentarily before the spark plug flashes, so knocking is usually perceived as a thud, ping, or...well, *knock* coming from the engine. Obviously knocking is something to be avoided since it not only works against the engine's motive power, but it's also tough on the mechanical parts.

At the very early stages of engine development, refiners discovered that the various types of gasoline blending components had different knock behavior. A consistent way to characterize when a blending component would knock turned out to be the *compression ratio*. In Figure 12-2 the compression ratio is shown as the cylinder's volume at the bottom of the stroke divided by the volume at the top of the stroke. The compression ratio at which self-ignition first takes place at the top of the stroke determines a blending component's octane number.

To make life easier, refiners devised a series of guide numbers to measure the compression ratio at which any gasoline component knocked. They defined iso-octane, C_8H_{18}, as 100 octane gasoline and normal heptane, C_7H_{16}, which knocks at a much lower compression ratio, as zero octane gasoline. By using a test engine, any gasoline component can be matched with blends of iso-octane and normal heptane.

Definition. The octane number of any gasoline blend or blending component equals the percent of iso-octane in the iso-octane/normal

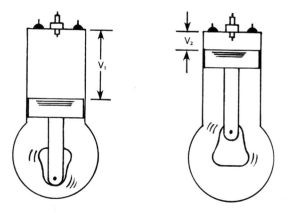

Fig. 12-2—Compression Ration Equals V$_1$: V$_2$

heptane blend that knocks at the same compression ratio as the gasoline or component being evaluated.

Testing for knock. An explanation of the test procedure might help. In its most basic form a test engine is used that has a top that can be screwed up or down to vary the compression ratio. The gasoline whose octane number is to be measured is fed into the engine while the head is being turned down. At some point, knocking will occur. Originally testers sat next to the engine with their heads cocked and listened for the knock by ear. Now they have a little more interesting life because they can use electronic detonation meters. After noting the compression ratio, the cylinder head is backed off. Two blends of iso-octane and normal heptane are concocted, one that the tester guesses will knock at a higher and one at a lower compression ratio. The octane numbers of these blends are known by definition (the % iso-octane). The tester runs each through the same test procedures noting the compression ratio when knocking occurs. By plotting the three data points, the octane number of the gasoline component can be read off a graph like that in Figure 12-3.

In that example, on a test engine a gasoline component knocked at a compression ratio of 8.1:1. Two test blends are made up, one with 88% iso-octane (88 octane) the other with 96% iso-octane (96 octane). In the test engine, they knock at 7.2:1 and 8.4:1, respectively. From the chart, the octane number of the gasoline component must be 94.0 octane.

Octane requirements. Now you know what octane numbers measure. Why are they important? The design of an engine demands that the fuel behave in a certain way. The compression ratio of an engine determines the amount of power it can deliver. The higher the compression ratio, the longer the power stroke, the more powerful the engine. Different size cars have different engine designs, therefore different requirements for gasolines of different octane numbers. Put more simply, you don't get to vary the compression ratio of your car by turning the head up or down. So you have to buy the quality of gasoline that accommodates the car and compression ratio you have.

Types of octane numbers. You need to know two more sets of nomenclature about octane numbers—the different kinds and their uses. First of all, the *tests* for octane numbers are run under two sets of conditions. The *research octane number (RON)* test simulates driving under mild, cruising conditions; the *motor octane number (MON)* test is run under more severe conditions and simulates operations under load or at high speeds. The

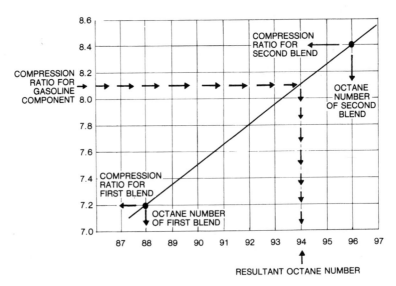

Fig. 12-3—Plotting the Octane Number Test

two measures, RON and MON, give an indication of a gasoline's performance under the full range of conditions.

In the late 1960s there was a controversy among the U.S. Federal Trade Commission (FTC), US refiners, and the car manufacturers about posting octane numbers on gasoline pumps. The FTC wanted octane numbers posted. The refiners had been advertising "100 octane gasoline" as their premium product, which referred of course to the RON. They wanted to post only the RON. They didn't want to confuse the public (or embarrass themselves) by putting the lower MON on the pump. The car manufacturers favored the MON as a better measure of how their products performed, and they were happy to sell cars that needed the "lower" MONs. The FTC considered posting RONs *and* MONs. "Too confusing," said the refiners and car manufacturers. The FTC finally arrived at a compromise by ordering the following to be posted on gasoline pumps:

$$\frac{RON + MON}{2}$$

That measure didn't have any particular meaning other than the fact the controversy was over. (R+M)/2 remains the nominal industry standard. The 87, 89, 91, or slightly higher and lower numbers you see on the gas pumps are the result.

The second piece of information about octane numbers relates to how they behave. When two gasoline components are mixed together, the octane numbers do not blend linearly. That is, the resulting octane is not the simple, volume-averaged octane. However, most fortunately, there is such a thing as a *blending octane number* for the RON and MON of every component that does blend linearly. The blending octane number is related to the true (test engine) number in a constant way, and is developed by experience. When references are made to RONs or MONs of components, they can mean either the true or blending octane number. To make things simple here, all references to octane numbers will mean the blending octane number, not the true number.

Blending for Octane Number. An example may help pull these ideas together. Take the blend of gasoline in the previous example where butane was added to achieve the vapor pressure spec. Calculate the RON and MON of the blend using the typical octane numbers in Table 12-3.

	Barrels	MON	RON
Straight run gasoline	4,000	61.6	66.4
Reformate	6,000	84.4	94.0
Hydrocrackate	1,000	73.7	75.5
Cat cracked gasoline	8,000	76.8	92.3
Normal butane	3,131	92.0	93.0
Total	22,131	78.1	87.4

Table 12-3—Blending for Octane Number

Now calculate how much alkylate must be added to meet a minimum specification of 84 MON and 90 RON, which would be 87 (R+M)/2. Alkylate has octane numbers of 95.9 MON and 97.3 RON.

	Volume	MON	RON
Gasoline blend	22,131	78.1	87.4
Alkylate	Y	95.9	97.3
Specification (minimum)		84.0	90.0

Table 12-4—Octane Numbers

In order to bring the 22,131 bbl blend with an MON of 78.1 up to an MON spec of 84, the following alkylate with MON of 95.9 must be added:

$$(22,131) \ (78.1) + Y \ (95.9) = (22,131 + Y) \ (84.0)$$
$$Y = 10,973 \ bbl$$

Meeting the MON spec doesn't mean meeting the RON spec. The same, but separate calculation is done to determine how much alkylate to add to meet the RON spec instead:

$$(22,131) \ (87.4) + Y \ (97.3) = (22,131 + Y) \ (90.0)$$

$$Y = 7,882 \; bbl$$

More barrels are needed to meet the MON spec; that decides the volume, because both the MON and the RON specs are minimums. The extra RON of the resulting blend, which turns out to be 90.7, is something refiners call *octane giveaway*.

You may have noticed a subtle problem undermines this example. If you add 10,973 bbl of alkylate to the blend to meet the minimum octane specifications, you no longer meet the RVP specification. By the use of two equations and two unknowns, *i.e.*, the volume of butane and the volume of alkylate, you could figure out the correct blend to satisfy both octane and vapor pressure specs. Maybe that's more algebra than you can stand right now.

Driveability. The composition of gasoline has another effect on how smoothly a car will run on a gasoline blend. As the gasoline ignites in an engine's cylinder, not all the molecules do or should burn at once. The piston needs a continuous push down the cylinder. A gasoline blend with a full spectrum of molecules usually provides the best chance of gradual, if you can use that word for something that happens during a thousandth of a second, burning. That gives great value to oil-based gasolines, which, as you saw in the chapter on distilling, are mixtures of all types of compounds. It also makes refiners cautious when they add very much chemical blending stock like ethanol which has only a single boiling point. It gives a flat spot on the distillation curve.

Leaded Gasoline

Until the 1970s refiners added lead to gasoline as a simple and economical way to increase the octane number. Lead, in the form of tetraethyl lead (TEL), increases the octane number of gasoline without effecting any other performance characteristics, including vapor pressure. Adding a small amount of lead, say about 3 grams of lead per gallon, could increase the octane number of a gasoline blend by as much as 5 RON. You can see that in Figure 12-4 from the effects on the various blending components.

Unfortunately, TEL is terribly toxic and in low concentration in vapor form can cause memory loss, blindness, or death. Because of this hazard, as late as the 1960s the Surgeon General of the United States set the maximum amount of TEL allowed in gasoline sold in the US at 4 grams per gallon. The US Environmental Protection Agency (EPA) succeeded to the authoritative position of the Surgeon General and when the pressure to do something about the poor air quality in many US cities became unbearable, the EPA mandated the use of catalytic mufflers in all new cars to improve the exhaust characteristics. It turns out that lead poisons the catalyst in these mufflers. To deal with both the health effects of airborne lead and the detrimental impact on the catalytic mufflers, the EPA then ordered a gradual phase out of the lead content in US gasoline, start-

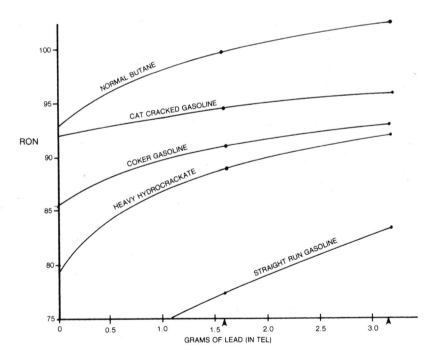

Fig. 12-4—**Effect of TEL in Gasoline**

ing in 1975. The use of lead in gasoline in America and most countries around the world is now just a memory, at least for those who never had a personal encounter with lead vapor.

Petrochemical Blending Components

With the advent of lead phase down, refiners began a desperate search for ways to maintain the octane level of their gasoline pool. They took the most obvious and readily available step—crank up the severity of the cat reformer, making higher octane reformate. Of course that reduced the volume of gasoline that could be produced, as you saw in the chapter on cat reforming. They faced the dilemma of building more refining capacity or looking for other ways to expand volume and octane at the same time. Oil company connections to the petrochemicals industry led them to experiment in the 1980s with a handful of readily available commodities: methanol, ethanol, TBA, MTBE, ETBE, TAME, and the totally unpronounceable THxME, THpME, THpEE and THxEE. Only some of them made the final cut. (Definitions explained later in chapter.)

	RVP, psi	RON	MON	(R + M)/2
Methanol	40	135	105	120
Ethanol	11	132	106	119
TBA	6	106	89	98
MTBE	9	118	101	110
ETBE	4	118	102	110
TAME	1.5	111	98	105

Table 12-5—**Blending Characteristics of Petrochemicals**

Methanol. One of the oldest industrial chemicals around is methanol, CH_3OH, also known as methyl alcohol or sometimes wood alcohol because of the historic practice of making it by chemically treating fresh-cut lumber from hardwood trees. Since 1923, chemical companies have used a more efficient process that starts with natural gas, or at least with its dominant constituent, methane. Some companies, in some parts of the world, use naphtha as a starter.

With a casual look at the formulas for methane and methanol, CH_4 and CH_3OH, you might think the route from one to the other would involve something like a catalyst and some water, maybe a little pressure and temperature. Unfortunately chemists have been searching to no avail for the silver bullet that would do the catalysis. They have found some, but none that have proved commercial. Instead, the route from methane to methanol involves a complicated, intermediate step, the creation of synthesis gas, a mixture of CO and hydrogen (H_2):

$$CH_4 + H_2O \longrightarrow CO + 3H_2$$
$$(Synthesis\ gas)$$
$$CO + 2H_2 \longrightarrow CH_3OH$$
$$(Methanol)$$

The formulas are deceptively simple. The process and hardware are complicated and pricey, and they involve expensive catalysts, and temperatures of 500-800°F and pressures of 4,000-5,000 psi. For that reason, almost all refiners chose to buy their methanol from chemical companies that specialized in the commodity and just blended it in.

Methanol has some properties that eventually undermined the enthusiasm for continuing to use it. First of all, it is corrosive and toxic and really bad if it gets on you. Second, it has an affinity for water in the same way your favorite bourbon or scotch whiskey does. Unlike oil/water mixtures, it won't settle out and has to be separated by distillation, which is not at all practical if water and methanol mix anywhere but in a refinery or chemical plant. To make matters worse, if methanol as a blendstock in gasoline picks up enough water, the methanol/water mixture will *phase separate*. It will remove itself from the rest of the gasoline blend. That makes handling difficult enough in refineries but especially in pipelines and terminals, which are not meant to be water free. (They didn't need to be because the water is removed at the terminals by drawing it off the bottoms of the tanks as *BS&W, bottom sediment and water*). Worst of all, if a methanol blend leaks out of an underground storage tank, which almost all gas stations have, the *hydrophilic* (water-loving) methanol can get into the underground water supply more easily than the hydrocarbon.

In the 1980s oil companies tried to commercialize a methanol blend in the US called M-85, made of 85% oil gasoline and 15% methanol. During the process they discovered more bad news about methanol. When it is burned in an internal combustion engine, it emits a scary carcinogen, formaldehyde, in the exhaust. Eventually the health hazards and water affinity caused M-85 and all other blends of methanol and gasoline to fail under their own weight.

TBA. At about the same time the chemical companies started promoting another chemical, tertiary butyl alcohol (TBA). TBA originated as a by-product from making propylene oxide. It can also be made on purpose by reacting butane with water or isobutylene and propylene. Refiners were attracted to TBA because it worked in sympathy with methanol as a *cosolvent*. TBA dissolved easily in both methanol and conventional gasoline blending components. So it helped methanol stay in solution in the gasoline blend, even as it picked up water. With the demise of M-85, most of the hype around TBA vaporized too.

MTBE. Methyl tertiary butyl ether (with a name like that, no wonder it's called MTBE) became an expensive but more user friendly successor to methanol. Like methanol, MTBE has affinity for water. But it is an ether, not an alcohol and it does not phase separate like methanol. Like methanol, it gives a big octane boost at 109 (R+M)/2 to the gasoline pool and has an RVP of only 8. (Both octane and RVP depend on what other components are blended in the gasoline.) Unlike methanol, it will not make you any sicker than the rest of the gasoline pool. On balance, refiners found MTBE attractive and started building their own MTBE plants.

The route to MTBE starts with iso-butylene and, ironically, methanol. Refiners still had an aversion to building methanol plants and continued to buy the commodity from chemical companies. Isobutylene was available from cat cracking and other cracking processes, but sometimes in limited supply. That created a market for iso-butylene and an incentive to build plants that would create iso-butylene from isobutane or even normal butane. See Chapter 16 for the process descriptions.

ETBE. For a while, refiners bought some ethyl tertiary butyl ether from chemical companies. Some have built ETBE plants, or used modified MTBE plants. ETBE has a higher octane number (109) and lower RVP (4) than MTBE, about the same water affinity and toxicity, but it is more expensive to make and not a big item on refiners' to-do lists.

Ethanol. To the surprise of traditional oil refiners came a strong contender, both politically and commercially, for winning the octane enhancer race in the 1980s and 1990s. Ethanol, or ethyl alcohol, comes from the natural fermentation of vegetables. For milleniums when farmers fermented grapes they got wine, potatoes they got vodka, and grain or corn they got whiskey. Ethyl alcohol, CH_3CH_2OH, is the operative ingredient in all these beverages. During the 1970s and 1980s, farm lobbies around the world successfully convinced their national or local governments that growing and fermenting corn, sugar cane, and some other crops and using it for automotive fuel would be good for their energy independence and their economies. Government subsidies brought on large amounts of natural ethanol capacity and created a consumer market for *gasohol*, a blend of 90% conventional motor gasoline and 10% ethanol.

Petrochemical companies also make ethanol from ethylene by reacting it with water in the presence of a phosphoric acid catalyst. The low conversion rate of 4-6% requires a lot of ethylene recycling through the reactor to get the job done. Consequently, ethanol from the petrochemical industry doesn't compete with ethanol from the agricultural sector, especially when government subsidies are in place.

Ethanol's octane number of 114 gives a real boost to the gasoline pool, but the RVP of 19 means that other lighter hydrocarbons have to be backed out of the pool to meet the vapor pressure limitations. That displaces all the butane from ethanol-enhanced gasoline and even some of the C_5's in the various blending components. Refiners have to run their fractionation columns differently and look for a home for the rejects. The pentanes are sometimes sent to the reformer to be converted to cyclic compounds or sold to chemical companies for ethylene plant feeds. The C_5 olefins can be made into the next group of exotics.

TAME, THxME, THpME, THxEE, and THpEE. This group of tongue strangling chemicals emerged when refiners needed to remove C_5, C_6, and C_7 olefins from the gasoline pools to meet the eventual specifications you'll read in a few paragraphs. Besides the olefins, the primary ingredient in these chemicals is either methanol or ethanol. TAME is tertiary amyl methyl ether (amyl is a synonym for a C_5, as in amylene, $C_5^=$.) THxME is tertiary hexyl methyl ether; THpME is tertiary heptyl methyl ether; THxEE is tertiary hexyl ethyl ether; and THpEE is tertiary heptyl ethyl ether. All these chemicals have high octane numbers, low RVPs, and no other bad stuff like sulfur in them. Refiners are loath to build these expensive plants unless regulators back them into a corner.

Combating Smog and Ozone

As governments developed more science about air pollution, they found more bad actors. They connected CO, oxides of nitrogen, including NO, NO_2, and NO_3, which they termed *NOx*, and the hydrocarbons floating around in the air, to the smothering ozone and thick layers of brownish smog that hung over the big cities. In the 1990s governments started mandating that gasoline must have a minimum oxygen content of several percent. Since hydrocarbons from crude oil have no oxygen content, that meant non-traditional blending components. Refiners had, by that time, already been blending the *oxygenates*, ethanol, MTBE, and some of the other ethers, all of which have an oxygen atom in each molecule. Now they had to accelerate the program to meet the government proscriptions.

The theory behind the mandate involved introducing oxygen to the combustion process to assure more complete burning of the hydrocarbons to CO_2 and H_2O, with virtually no CO. It also reduced the amount of unburned hydrocarbons coming out of the engines. The change was intended to enhance the operation of the catalytic mufflers (that were already supposed to be doing the same thing) and improve those cars that had old-fashioned mufflers.

TOX, NO$_x$, VOCs, and SO$_x$

This unsavory group, which sounds like the name of a law firm formed by the children of itinerant circus performers, was the next show that came into town. Extensive, invasive engineering research found these culprits spewing out the tailpipes of cars, vaporizing out the gas tank fill pipes, and generally leaking throughout the distribution systems. Governments, oil companies, and environmentalists adopted a shorthand clarion call, TOX, NO$_x$, VOCs, and SO$_x$, which stand for toxic compounds, nitrogen oxides, volatile organic compounds, and sulfur oxides.

The conventional wisdom deemed a number of givens. The heaviest parts of the gasoline blending components, those at the high end points, contributed to unburned hydrocarbon getting through the combustion system. The lightest of the hydrocarbons such as butane were evaporating or leaking out of the gas tanks, the engine seals, and the gas pumps before they could even get to the combustion chamber. Benzene in small concentrations was positively identified as a carcinogenic threat and therefore one of the toxic compounds. Other aromatics compounds could result in unburned emissions, including benzene. NO$_x$ and VOCs reacted with sunlight to create smog and ozone. SO$_x$ caused the catalyst in mufflers not to work so well, emitting more VOCs and CO. And on and on and on.

The consequence of all this good news/bad news was a succession of rules for the content and behavior of gasoline under the banner of *refor-*

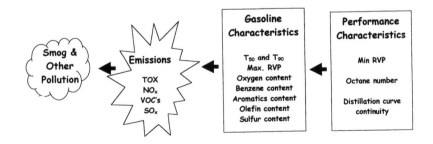

***Fig. 12-5*—Blending Gasoline to Meet Environmental Specs**

mulated gasoline for fuel sold in densely populated areas where ambient conditions don't disperse pollutants very effectively.

The complexity of gasoline blending now reaches art form, as you can see in Figure 12-5. The object of the regulations is to prevent air pollution, but TOX, NO_x, VOCs, and SO_x coming out of cars contribute to different forms of pollution in mysterious, non-linear ways. The emissions from burning gasoline cannot exceed various combinations of the four *emitants* (a term emerging from the pollution-speak community) in panel two. The control mechanism sets limits on combinations of the four.

At the same time the generation of the four emitants is connected in complicated ways to the gasoline characteristics in the third panel. T_{50} and T_{90} (the temperatures at which 50 and 90% of the gasoline boil off) and RVP measure volatility and shape of the distillation curve. The others relate to the composition. Because of their impact on the four emitants, all these properties have explicit limits, some by statute, at least in the US.

Finally, in order for gasoline to work well in car engines, refiners must make gasoline that meets their own performance specifications, shown in the fourth panel. (Distillation curve continuity is the property refiners watch to assure the gasoline blend has a full range of compounds so it ignites smoothly over the whole power thrust of the piston. That of course limits the addition of single boiling temperature blend stocks like the alcohols and ethers).

Gasoline Blending and its Impact on Operations

Do refiners have a clue how to deal with all these constraints? How do governments know if they are complying? There is no pretending that optimizing the blending of gasoline is simple. Consider for a moment the ever-increasing levels of complexity as refiners set the operating conditions in the processing units:

- Given the demands for three grades of gasoline and the availability of components, blend up the requirements—with no leftovers

• Now consider varying the operating conditions of some of the process units. Change the severity on the reformers to adjust yields versus octane numbers, but be mindful of the benzene and aromatics content; increase the temperature in the cat cracker to generate more olefins and ultimately more alkylate, etc

• Finally, consider diverting streams in and out of units. Send cat cracked light gas oil to be blended to furnace oil rather than hydrocracked; remove the light ends from straight run gasoline to reduce RVP; cut the benzene precursors out of reformer feed; cut the bottoms off the straight-run naphtha (reformer feed) to make more kerosene/turbine fuel, and so on.

Refiners have put in place predictive computer models of increasing complexity and are developing others to help cope with these challenges. The models involve the old techniques, linear programming to simulate refinery operations. The intakes, outturns, capacities, and costs of each operation, from distilling to blending, are described using equations and numerical values. The crude oil availabilities and costs, and the product demands and prices are specified. The linear programming technique will find the solution to the equations (there are usually many) that makes the most profit.

The computer is necessary because of the thousands of calculations that are necessary to find the optimal solution. Even this is only an approximation, for several reasons:

• The data fed into the models are estimates of the process unit yields. Depending on any number of things (time since the last shutdown, catalyst activity, air temperature, cooling water temperature, etc.) the yields and octane numbers could vary

• The crude composition could vary

• The demands and prices could vary

Further, the inevitable unscheduled shutdowns in some part of the refinery will interrupt orderly flow. Nonetheless, as an analytical technique to develop a model or a plan, the linear program is an invaluable tool.

To get from panel three to four in Figure 12-5, refiners are now adding other mathematical techniques like regression analysis, followed by lots of testing to see if the theoretical results came out correctly.

Conclusion

The subject of gasoline blending brings into focus most of the operations in a refinery. The rudiments of meeting octane and RVP specs are simple enough, but optimizing gasoline blending implies optimizing the entire refinery.

As a gesture of conciliation to you after reading this exhausting litany of refiners' problems, no TOX, NO_x, VOCs, and SO_x blending problem will be included in the exercises.

Exercises

1. Define the following terms:

vapor pressure	oxygenate
RVP	compression ratio
power stroke	RON and MON
vapor lock	leaded gasoline
pressuring agent	octane enhancement
knocking	gasohol
driveability	TOX, NO_x, VOCs, and SO_x

2. a. Calculate the amount of normal butane needed to produce a 10.0 psi RVP for a mixture of 3,000 barrels of alkylate, 2,500 barrels of reformate, 6,000 barrels of hydrocrackate, and 3,600 barrels of cat cracked gasoline.

   b. How much MTBE (with MON of 103 and RON of 115) needs to be added to bring the blend up to 84 MON and 90 RON?

   c. What problem with the blend spec occurs when you added the MTBE?

   d. Why does the driveability get affected if you add too much MTBE?

Distillate Fuels

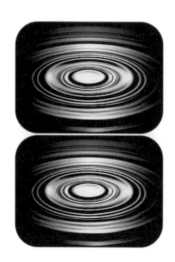

Warmth, warmth, more warmth! For we are dying of cold and not darkness. It is not the night that kills, but the frost.

-The Tragic Sense of Life,
Miguel de Unamuno

Diesel fuel and furnace oil are the two fuels refiners make from the light gas oil streams. You might say they are the one fuel made because in some places diesel fuel and furnace oil are the same product, except for some additives that might be put in just before they are sold.

Diesel Engines

A cutaway view of a diesel engine looks a lot like a gasoline engine. The one difference you might notice is the absence of a spark plug. The mechanics of a diesel engine depend on the fuel self-igniting, the very thing important to avoid in a gasoline engine. (You can see a diesel engine cutaway view in Figure 14-3 in the chapter about residual fuels).

The control of self-ignition depends on very acute timing. Unlike gasoline engines, air is not mixed with the fuel before it is injected into the cylinder in the intake stroke. Only air is injected initially. As the piston moves to the top of the compression stroke, the air gets hotter and hotter as it compresses. Just as the piston reaches the top of the stroke, the diesel fuel is injected into the cylinder. On contact with the superheated air, it ignites and causes the piston to begin the power stroke.

Several distinct phases happen as the fuel enters the cylinder. First, the fuel arrives in liquid form, although it may be sprayed to disperse it. As the liquid hits the superheated air, some of it vaporizes and rises to the self-ignition temperature. The second phase begins as combustion occurs and starts igniting the surrounding vapor and vaporizing liquid. The power stroke begins. Finally as the rest of the liquid is pumped into the cylinder, it also ignites, maintaining or increasing the pressure on the piston. All this happens in less than a thousandth of a second.

The precision required of the timing mechanism becomes apparent. If the fuel is not injected at the correct moment and rate, too much fuel could ignite at once, causing an explosive power stroke rather than a controlled surge. Similarly, the fuel itself must have the combustion characteristics to burn within narrow limits-not so fast that it is explosive, not so slow that some hydrocarbons are left unburned at the beginning of the exhaust phase.

Different diesel engines are designed to handle a wide variety of fuels, from distillates to residual fuels. This chapter covers only the distillate fuels. In the next one, chapter 14, Asphalt and Residual Fuel, you'll find the technology of burning heavy fuels in diesel engines. That will introduce the two cycle engine, one that is very popular in marine propulsion, and the fuels that feed it. The rest of this chapter will deal only with distillate fuels.

Diesel Fuel

To understand the art of making diesel fuel, you have to forget half what you learned about blending gasoline. In addition, diesel fuel quality has to accommodate three indistinct market segments, automotive, heavy truck, and off-highway uses. This last category includes construction equipment, irrigation pumps, electricity generators, railroad engines, and the like.

Cetane number. For gasoline engines, self-ignition is an abomination. For diesel engines, it is the essence. Refiners use cetane number as the measure of the self-igniting quality of a diesel fuel. Reminiscent of octane number and its derivation, the cetane number of a diesel fuel is determined using a test fluid, a mixture of cetane ($C_{16}H_{36}$) and alpha-methyl-naphthalene ($C_{11}H_{12}$). The cetane number equals the percent $C_{16}H_{36}$ in a mixture that self-ignites just like the diesel fuel being evaluated.

Car manufacturers have designed their diesel engines to demand a higher cetane number diesel than the truck manufacturers. "Gas stations" normally sell 45 cetane diesel fuel. At a truck stop you'll find 40 cetane diesel. The off-highway diesel equipment is more like heavy truck diesels than car engines and will typically use 40 cetane diesel. Some oil companies sell a premium diesel at gas stations, which runs 47-49 cetane. They often will put an additive package in premium diesel to help keep the engine clean and to help justify the extra cents they charge for the word premium.

The PONA content has an important impact on the cetane number of diesel fuel, just as it does on the octane number of gasoline. In a gasoline engine, aromatic compounds resist self-ignition. They have high octane numbers. Paraffins have low self-ignition thresholds and low octane numbers. In diesel engines, the low threshold makes paraffinic fuels more desirable—they readily self-ignite and have higher cetane numbers. The aromatics (prevalent in cracked gas oils) are less attractive and have lower cetane number because they resist self-ignition.

Sulfur content. In America, compared to the heavy impact of environmental concerns on automotive gasoline, restrictions on automotive diesel have been only modest. Regulators have focused only on the sulfur content of diesel used on the highways in trucks and cars, limiting it recently to 0.05%. Off-road users can still use the same quality of fuel that oil companies sell as furnace oil, which is often 1–2% sulfur.

Pour point. To add to the confusion about cetane numbers and sulfur, refiners have to deal with the pour point of diesel. When it gets really cold, gas oil can start to crystallize. The long chain paraffins are the culprits since they form solid wax particles, enough to stop the fuels from moving through a fuel line.

Pour point is defined as 5°F higher than the temperature at which a liquid stops flowing.

A companion property to pour point is the *cloud point*, the temperature at which the wax crystals start to form. Diesel engines all have protective screens at the end of the fuel lines to prevent any extraneous "stuff" from entering the atomizer, which sprays the fuel into the cylinder. Wax crystals can clog this very fine mesh, causing the engine to sputter or stop.

In the winter, refiners will have to add kerosene to diesel fuel to keep the pour point and cloud point below the ambient temperatures of the region. The good news is that kerosene enhances the cetane number of the diesel fuel because kerosene is generally more paraffinic than gas oil. The bad news is that kerosene, normally used to make jet fuel, sells for more than diesel.

Blending components. All the light gas oils are candidates for diesel fuel blending, but some are better than others. Straight run light gas oil is generally very paraffinic and usually has a cetane number in the 50-55 range. The cracked light gas oils from the cat cracker, coker, or thermal cracker generally have high olefinic, aromatic and naphthenic concentrations and are in the 32-35 cetane range. Gas oil from the hydrocracker, which has pumped a lot of hydrogen into the products, tends to be high on paraffins and naphthenes, low on aromatics, and zero olefins, and so makes a good diesel blending component. Kerosene has a cetane number of about 55.

Ideally, refiners would make their diesel fuel from straight run and hydrocracked light gas oil. They would keep all the cat cracked light gas oil out of diesel and use it for furnace oil, which has no cetane number requirement. They could also use it as feed to a hydrocracker to make gasoline. But refiners don't always have the markets for furnace oil or a hydrocracker and some, if not all cat light gas oil, ends up in diesel blends. Sometimes they have to put *ignition accelerator additives* or kerosene in to get the cetane number up to specification. For the typical commercial additive available today, the addition of 0.5% can lift the cetane number by as much as 10 cetane numbers. The additives are expensive and refiners avoid using them if they can get away with it.

Furnace Oil

The gas oil range hydrocarbons have a number of physical characteristics that make them a popular fuel for home heating:

• They carry more heat per gallon than the lighter hydrocarbons such as kerosene, naphtha, or LPG

• They are cheaper to transport than LPG, and in many cases natural gas, because they do not have to be kept under pressure

• They do not pose the special safety hazards of natural gas and LPG

• They are easier to transport and burn than residual fuels because they do not have to be heated to move around

• They usually have lower sulfur contents than the heavier residual fuels, and if they don't, they can be desulfurized with simpler equipment (see the discussion about hydrotreaters in chapter 15)

Two Oil and All That

The number system for fuel oils is almost lost in past records.
Originally refiners established six grades of fuel oils which
more or less related to their cut points and to the application.

- *Number 1 oil was in the kerosene range and was used for lighting, cooking, and heating*
- *Number 2 oil in the light gas oil range was used as domestic heating oil, but too heavy for lighting and cooking*
- *Number 3 oil was an intermediate grade that died an early death*
- *Number 4 oil was a mixture of residual fuel and light gas oils that could be used as marine diesel without heating*
- *Number 5 oil was a commercial heating oil and marine diesel that required heating to move and burn it. It has also been rebranded extensively to road oil, poured directly on dirty roads for a semi-permanent hard surface that helps keep the dust down*
- *Number 6 oil is a heavy fuel oil used as boiler fuel in industry and ship's bunkers and has to be heated to move it*

For all these reasons, furnace oil has been used extensively for residential and commercial heating, although less in the US than other countries because of the abundance of cheap natural gas.

Furnace oil has a number of synonyms—number 2 fuel, distillate fuel, two oil, heating oil, home heating oil, and outside the US, gas oil.

Specifications for furnace oil cater to the design limitations of residential heating systems. *Flash point* is important for safety; pour point protects against extremely cold weather, which is when you want to use furnace oil the most.

Flash point. The lowest temperature at which enough vapors are given off to form a combustible mixture is called the flash point. It therefore accounts for both volatility and flammability. For furnace oil, flash point is set at a limit that protects against vapors leaving the storage tank and finding another source of ignition. (Many liability suits have historically come out of accidentally mixing gasoline with furnace oil during transport. The gasoline vapors leak out of basement heating oil storage tanks, find the pilot light on a nearby hot water heater, and explode).

Blending furnace oil. Furnace oil has no self-ignition issues and no cetane number requirements. The main objective of burning furnace oil is, of course, generating heat-all the gas oil range streams have about the same heat content. Consequently, refiners and marketers can make diesel fuel and re-brand or, in effect, downgrade it to furnace oil right out of the same tank.

Exercises

1. How do gasoline and diesel fuel differ in the following characteristics?

   • How ignition occurs
   • How octane and cetane numbers are measured
   • Components that favor octane and cetane numbers
   • Components that disfavor octane and cetane numbers

2. What's the difference between furnace oil and diesel fuel?

14

Asphalt and Residual Fuel

Muddy, ill-seeming, thick bereft of beauty.

-The Taming of the Shrew,
Shakespeare

A refiner has two market alternatives to handle the bottom of the barrel—residual fuel and asphalt. The choice could depend on the quality of various crude oils available and on access to the markets.

Asphalt

Sitting in the very bottom of the barrel of most crudes are varying amounts of *asphaltenes*, which are very complex molecules. Asphaltenes are polyaromatic compounds with high carbon to hydrogen ratios because of the way the aromatic rings connect to each other. Organic chemists sometimes nudge each other, giggle, and refer to them as trimethyl chicken wire, as they think about the picture in Figure 14-1. The number of carbons in the molecule usually exceeds 50. With a high enough asphaltene content, the bottom of the barrel material makes a good, strong, stable binder—asphalt.

Asphalt can be categorized into the four groups shown in Figure 14-2, although the categories overlap. There are straight run asphalts, blown asphalts, cutbacks, and emulsions. Not all crude oils make good asphalt. Some crudes make good blown asphalt, others might make good emulsions. Refiners usually start an evaluation in their laboratories, but then make test runs in their plants to determine if candidate crudes are asphalt suitable.

Straight Run Asphalts

Deep flashing of the straight run residue of an asphaltic crude will yield flasher bottoms that can be used directly as asphalt. The bottoms are a black or dark brown material, very viscous (thick) or solid at ambient temperatures. The higher the flasher temperature, the more viscous the asphalt.

There are two usual tests for grading this type of asphalt—softening point and penetration. The *softening point* occurs at the lowest temperature at which an object with a standardized weight and shape will start to sink into the asphalt. The most popular test method uses a steel ball. Commercial grades of asphalt have softening points somewhere around 80-340°F.

The hardness of an asphalt, once it's applied, is measured by its *penetration*. The test apparatus for penetration has a long needle with a weight on top. The depth to which the needle penetrates the asphalt over a standard period of time at a given temperature is the measure of penetration. Very hard asphalts are zero penetration, while the softer ones range up to 250.

A wide variety of grades of asphalt can be dispensed at a refinery by making *blocked-out runs*. For a period of time, the flasher can be run to fill a tank with 40-50 penetration asphalt. Operating conditions on the flasher can then be changed to make a 250 penetration grade, running the bottoms to another tank. By blending these two grades directly into trucks, tank cars, ships, or barges, any grade in between 40 and 250 penetration can be supplied.

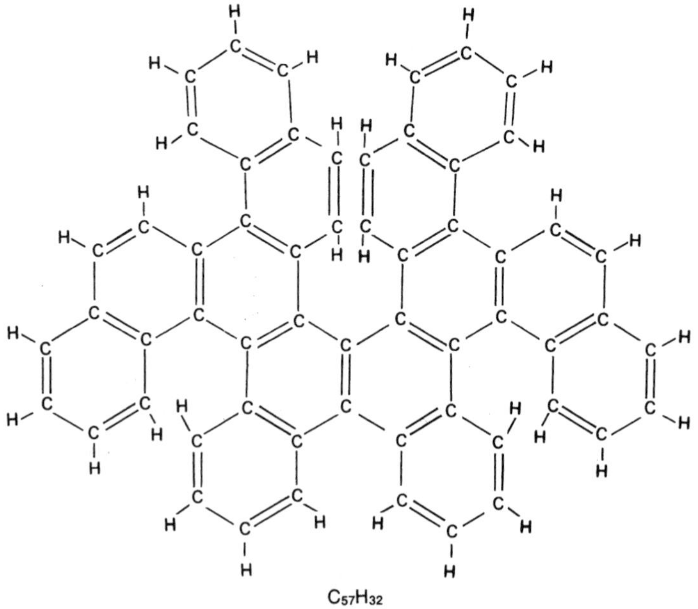

$C_{57}H_{32}$
An Asphaltene
(Artist's concept)

Fig 14-1—An Asphaltene

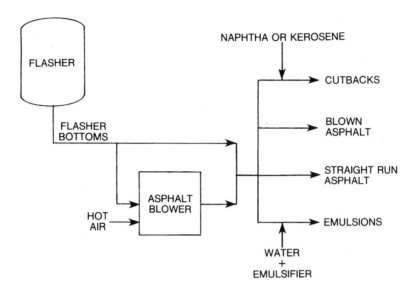

*Fig 14-2—*Asphalt Options

Blown Asphalt

This is not slang for the result of a refiner making off-spec asphalt. Rather, as an alternative to producing asphalt via vacuum flashing, a refiner can chemically change the consistency of the softer grades in an *asphalt blower*. This unit blows hot air into flasher bottoms, causing a chemical reaction. Either the oxygen bonds to the asphalt, or hydrogen atoms from the flasher bottoms combine with the oxygen to make water, which evaporates—or both. A harder, more rubbery asphalt results. The lower penetration asphalts can be achieved via this route, which is sometimes cheaper due to the composition of the pitch content of a particular crude.

Cutbacks

The application of asphalt requires that the material flow at temperatures not highly elevated. In road construction, e.g., the asphalt must completely surround the aggregate (gravel) and sand with which it is mixed. Normally, asphalt is mixed and applied at heated temperatures. Prior to the acknowledgement of the environmental impacts of asphalt handling, refiners used to sell asphalt with a thinner such as naphtha or kerosene added to make the handling easier at the construction sites. The thinner, or diluent, softened the asphalt and permitted lower application temperatures. After the asphalt was applied, the diluent evaporated, leaving the hard, durable asphalt behind. That all seems out of the question now.

Refiners were happy to see the demise of cutback asphalts. In addition to the burden of pollution, the thought nagged them that they were putting valuable hydrocarbons in the asphalt only to have them evaporate into thin, and sometimes not so thin, air. What a waste. As the environmental pressure built, they created a more durable and planet-friendly asphalt, emulsions.

Emulsions

Emulsions achieve the same flexibility in field equipment but avoid the environmental problems. Emulsions are a mixture of about 50-70% asphalt and 30-50% water. Of course asphalt, which is an oil, and water don't mix very well by themselves. To get around that problem asphalt blenders use an emulsifying agent, a solution that looks like, and behaves much like soap. The molecules of the emulsifier have one end that is *hydrophilic*, i.e., have an affinity for water. The other end of the molecule is *oleophilic*, i.e., attracted to oil. Honest! Dishwashing soap works the same way to remove the grease from your dinner plates. After a while there's so much grease attached to the oleophilic ends of the soap, the dishwater won't remove any more from the plates.

A little emulsifying agent goes a long way. With about 1%, the asphalt and water will stay mixed during transport and application. After the asphalt emulsion has been laid, the water evaporates, leaving behind the asphalt and the little bit of emulsifier.

Emulsion asphalts used for road construction are usually the softer grades, 200-300 penetration. Industrial asphalts for coatings generally use harder grades—40-50 penetration. Often clay is mixed with the asphalts for roofing (shingles, tarpaper, and felt) and flooring materials (damp-proofing coatings and tiles).

Some clever researcher discovered asphalt particles in emulsions are by nature negatively charged—called *anionic emulsions*. Many aggregates used as mixes with asphalt are also by nature negatively charged. Sometimes it is difficult to completely coat the negatively charged aggregate with the negatively charged asphalt emulsion, especially when they are mixed in damp weather when electricity can really flow. To combat this problem, special emulsions have been developed in which the asphalt particles are positively charged. These emulsions, called *cationics*, readily grab the negatively charged aggregates. Asphalt emulsions used for road construction are now mostly cationics.

Residual Fuels

Residual fuel is what's left after you make everything you want. That's of course why it's called *residual*. In many non-US refineries, residual fuel is no more than the crude distilling unit bottoms, called *long residue*. Most US refineries use long residue as feed to a flasher. In that case, flasher bottoms (pitch) are the primary ingredients of *resid*.

Residual has always commanded lower prices than all other fuels for two reasons; one physical and the other commercial. Moving and consuming resid requires special equipment. The pipelines, transportation, and storage must be heated so the residual fuel doesn't solidify. Secondly but more importantly, almost all resid is a leftover from producing the other product. Its short term supply elasticity is nearly zero. *Resid disposal* is the term buyers love to hear when refiners talk about selling their residual fuel.

Residual fuel markets. The lower price of residual fuels inevitably attracted users to find ways to use them. After all, residual fuels have more heat per gallon than any of the other fuels, as you see if you read through chapter 22 on Fuel Values and Heating Values. The traditional use for residual fuels was boiler fuel, the simplest possible way to burn hydrocarbon. The heated residual fuel is sprayed into a large boiler where the constant fire vaporizes the droplets and burns them. A maze of tubes in the boiler carries water, which is heated to steam or even superheated steam. This steam then drives electricity generators, ships' propulsion systems, or other mechanical apparatus.

Eventually the mechanical engineers developed diesel engines that could use the heavier, cheaper residual fuels in diesel engines. Diesels had a mechanical advantage over boilers, since more of the energy generated during combustion gets harnessed and doesn't go out the stack (used to be called the smoke stack, but you can't say that any more) and get lost.

The popular form of diesel for industrial applications, again the electricity generators, ships' propulsion, and other heavy duty uses, is the two cycle engine. These mammoth machines have 8-12 pistons and cylinders 2-3 feet in diameter with a stroke 5-8 feet long. They operate at much slower rates than automotive diesels, perhaps 100-135 rpm.

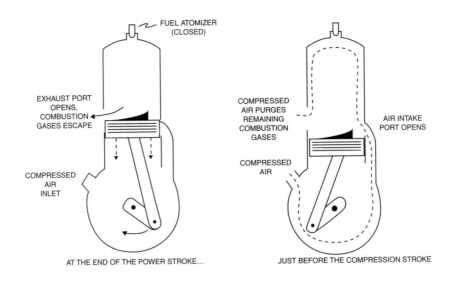

Fig 14-3—Two Cycle Diesel Engine

You may wonder how a two cycle engine can get all the things done that a four cycle engine does. Why aren't all the engines two cycle if they can do the job of four? If you haven't wondered or don't care, skip the next few paragraphs. The two cycle engine shown in Figure 14-3 has the same compression and power stroke as the four cycle diesel engine. The other two cycles, intake and exhaust, are handled in a split second operation in between compression and power strokes.

As the piston approaches the bottom of the power stroke, the exhaust port opens and the combustion gases whoosh out. Milliseconds later, the inlet port opens and compressed air shoots in, *scavenging* or cleaning out remaining combustion gases and filling the cylinder with fresh air. Both ports close and the compression stroke begins. The escaping combustion gases do some more work outside the engine when they enter a turbocharger. That drives the fresh air compressor and contributes to the efficiency of the two cycle engine.

Because of the precise timing of the intake and exhaust, two cycle engines need to operate at slower rates than four cycle engines. In addition, the burn rates of the heavy fuels are slower than distillate fuels. The slow burn rate also means a longer stroke. Two cycle engines are inherently more polluting than four cycle because they exhaust the combustion gases so quickly, with some unburned hydrocarbons likely being left as the pressure drops to exhaust them.

Blending resid. Like every other product on the refiner's slate, residual fuel has changed in recent decades in response to environmental concerns. The sulfur content of residual fuel varies, depending on what country, and even what city it will be burned in. Resid is often used in electric power plants that have *stack gas scrubbers* in place. These devices capture most of the sulfur that had been in the residual fuel but make SO_xs when the resid burns. Stack gas scrubbers are expensive and don't compete well with refinery sulfur removal facilities or resid destruction units in refineries, and so high sulfur resid use in power plants continues to decline around the world.

Because resid is a by-product, the other specifications for marketable products are very loose. The most important spec is viscosity, a measure of how much a fluid resists flowing, as shown in Figure 14-4. (Shampoo is more viscous than water). The standard unit of measure for viscosity is *centistokes*. Since viscosity changes with temperature (your shampoo will flow a lot more easily in a hot shower than if you use it in a cold lake), the centistokes are measured at (usually) either 80°C or 100°C. Flasher bottoms generally need to have some kind of a diluent or *cutter stock* added to meet the maximum viscosity spec. Typically, the diluent is a heavy stream of low viscosity and has the lowest value around the refinery as feed to a thermal cracker, coker, or hydrocracker. Cat cracked heavy gas oil is typically used, if available.

Like distillate fuels, the flash point of residual fuels sometimes binds. Because residual fuels have to be heated to be pumped, the flash point is even more critical than for distillates. At the same time, residual fuels tend to be a garbage dump where a lot of miscellaneous offspec or otherwise bad streams can be hidden. Flash point can often limit what refiners can bury in resid.

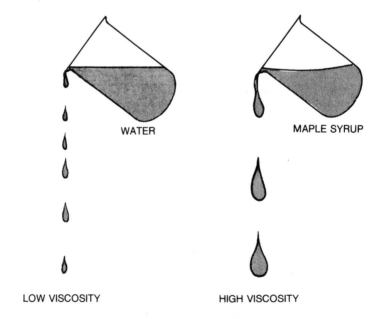

WATER

MAPLE SYRUP

LOW VISCOSITY

HIGH VISCOSITY

*Fig 14-4—***Viscosity is Resistance to Flow**

Caribbean refiners originally instituted a technique for increasing the sales of residual fuels made from high sulfur crudes. The hydrotreaters described in chapter 15 can remove sulfur from heavy and light gas oil range streams relatively inexpensively. By blending the high sulfur flasher bottoms with enough desulfurized gas oils, the resulting residual can meet a lower sulfur spec. Some people referred to this as "dilution, the solution to pollution."

Exercises

1. What new class of compounds was introduced in this chapter that makes asphalt possible?

2. Why doesn't every refinery make asphalt instead of residual fuel?

3. What two designs can shipbuilders use to power a ship with residual fuel?

chapter fifteen

Hydrogen, Hydrotreating and Sulfur Plants

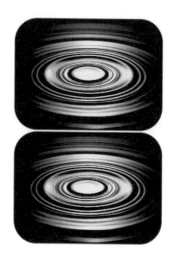

Neither to smell rank nor to smell sweet pleases me.

-Epigrams, Decimus Magnus Ausonius

C rude oils have all sorts of contaminants. As the petroleum fractions travel through the refinery processing units, these impurities can have detrimental effects on the equipment, and the catalysts. When they come out the refinery gate in finished products, they can pollute the environment. Consequently governments or the refiners themselves may impose limits on the content of some impurities, like sulfur.

Hydrotreating does an effective job of removing many of the contaminants from many of the streams. Hydrogen (H_2) is the operative input to the hydrotreating process.

Hydrotreating

Refinery streams that have C_6 and heavier hydrocarbons in them are likely to have some sulfur hidden in the mixture in the form of obscure compounds, even if they come from sweet crude. The sulfur atoms can be attached or imbedded anywhere in various molecules and therefore mechanical processes like distillation can't remove it. It has to be done chemically. Hydrotreating is the normal process used to break the sulfur away.

In the hydrotreating process, the stream is mixed with H_2 and heated to 500-800°F. The oil/H_2 stream is then charged to a vessel filled with a catalyst in pellet form (Fig. 15-1). The catalyst promotes several reactions:

- The hydrogen combines with the sulfur atoms to form H_2S
- Some nitrogen compounds are converted to ammonia
- Metals entrained in the oil deposit on the catalyst
- Some of the olefins and aromatics get saturated with hydrogen
- As the contaminants crack away from the hydrocarbon, some methane, ethane, propane, and butanes form

The most popular catalysts being used these days are cobalt and molybdenum oxides on alumina. These exotic metals have proven to be resistant to poisoning from contaminants and easier to regenerate. They also are highly selective, which means they don't have a lot of other side reactions going on besides the removal of sulfur and nitrogen.

The stream coming from the reactor is sent to a flash tank where most of the propane and lighter, including the unused H_2, the H_2S, and the tiny bit of ammonia, go overhead. To completely strip out these light ends, a small fractionator is generally tacked onto the tail end of the process.

The importance of hydrotreating has been gradually increasing for many years for three reasons:

- Pretreaters in front of the catalysts in cat reformers, cat crackers, and hydrocrackers protect the increasingly sensitive catalysts from getting fouled by sulfur and some metals
- Governments are continually lowering the allowable sulfur content in gasoline and fuel oils
- Hydrotreating will change the PONA composition of fuels, reducing the olefins and aromatics, increasing the paraffins and naphthenes. Sometimes this cuts some slack in other parts of the refinery. They might have been constrained by having to meet olefins and aromatics limits in the finished products

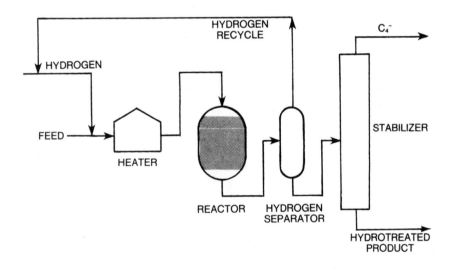

Fig 15-1—Hydrotreater

Some specific examples of all three of these drivers follow.

Jet fuel hydrotreating. Hydrotreating will improve the burning characteristics of distillates, especially jet fuel. The kerosene fraction can contain a large percentage of aromatic compounds that have higher carbon-to-hydrogen ratios. When these compounds burn, the deficiency of hydrogen can cause smoke. As a matter of fact, one of the specifications on jet fuel is the *smoke point*. Smoke is nothing more than unburned hydrocarbon, molecules that didn't make it all the way to H_2O and CO_2 (see chapter 22 on fuel values and burning stuff). Today you can sometimes see a darkish trail behind jet passenger planes as they rise from take-off at maximum power, a mode in which unburned hydrocarbons are likely to make it through the jet engines.

The apparatus used to measure smoke point is similar to a kerosene lantern. A reservoir of fuel is fitted with a wick that can be cranked up or down to vary the length and the flame size. Size matters—the smoke point is a measure of how far the wick can be cranked up before smoke is visible above the flame. Smoke point equals the wick length, in millimeters.

Kerosene with a low smoke point can be improved by hydrotreating. During the process, the aromatic rings get saturated with hydrogen, converting them to naphthenes with more hydrogen-to-carbon ratios that burner cleaner.

Distillates hydrotreating. Distillate and jet fuel hydrotreater can and often are used interchangeably. Distillates used for diesel fuel need at least some treating to get the sulfur levels on-spec. In addition, hydrotreating the cracked light gas oils reduces the amount of aromatics compounds and raises the cetane number.

Pyrolysis gas hydrotreating. One of the co-products of making ethylene from naphtha or gas oil is pyrolysis gasoline (covered in chapter 19). This gasoline stream has a high content of aromatics, olefins and *di-olefins*—all molecules that have one or more sets of double bonded carbons. Often, pyrolysis gas is only suitable for gasoline blending in small concentrations. It smells bad, has a funny color, and forms gum in fuel injectors.

Hydrotreating pygas saturates double bonds, eliminating most of the undesirable characteristics. The octane rating may decline during hydrotreating as some of the aromatics and olefins get converted to naphthenes and normal paraffins.

Cat feed and reformer feed hydrotreating. To protect the catalyst and improve conversion rates, cat reformer feed is almost always hydrotreated. Hydrotreaters in front of cat crackers have a triple purpose. They protect the catalyst, they result in cleaner CCU products, and the conversion of some of the aromatics in the cat feed enables better conversion rates. In this case, the hydrotreating catalyst doesn't just protect the CCU catalyst; it works in sympathy with it.

Residual hydrotreating. Residual fuels are likewise coming under environmental pressure, and, somewhat belatedly, residual desulfurization facilities have been commercially developed. Although the flow diagrams are similar to lighter stream hydrotreaters, the hardware and yields are different. Residual streams have much lower hydrogen -to-carbon ratios. In addition to the larger presence of excess hydrogen in the reactor, high pressure must be maintained to minimize coke formation. So a residual hydrotreater must be built as sturdily as a hydrocracker—an expensive proposition.

The yield from hydrotreating residuals has higher light end production. In the large molecules prevalent in residual fuels, especially the "trimethyl chicken wire" compounds, sulfur, nitrogen, and metals cannot be removed without literally destroying the molecule to spring the imbedded pollutant. In the process the smaller molecules result.

Hydrogen Plant

The normal source of hydrogen in a refinery is the cat reformer. The light ends of the reformer column contain a high ratio of H_2 to CH_4, so the stream is de-ethanized or depropanized to get a high concentration of H_2 stream.

Sometimes the reformer H_2 cannot satisfy all the H_2 requirements in a refinery. This is especially true if there is an H_2-hungry hydrocracker in operation. H_2 can be produced on purpose in a plant called a *steam methane reformer* (SMR) as shown in Figure 15-2.

As the design engineers considered ways to make hydrogen, they looked for chemical compounds with a high proportion of hydrogen in order to waste as little as possible of the remaining material or the energy used to process it. The two compounds they finally chose were almost too obvious, CH_4 and H_2O.

The trick in the SMR is to spring as much of the hydrogen from CH_4 and H_2O as possible, but with a minimum amount of energy (fuel) consumed in the process. With the aid of some very useful catalysts, some high temperatures and pressures, and heavy duty equipment, the SMR operates in four stages:

1. *Reforming.* Methane and steam (the H_2O) are mixed and passed over a catalyst at 1,500°F, resulting in the formation of carbon monoxide and hydrogen:

$$CH_4 + H_2O \longrightarrow CO + 3H_2$$

But in a refinery, not much use can be found for CO. That may not be so in a chemical plant, where they call this combination of CO and H_2 gases synthesis gas and use it to make some alcohols and aldehydes. In a refinery they take a next step.

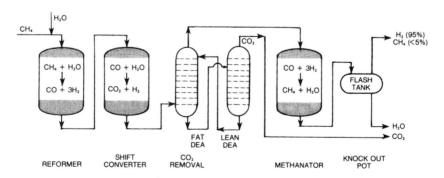

Fig 15-2—Steam Methane Reformer

2. *Shift conversion.* Not content with the H_2 already formed, the process exploits the CO. More steam is added over another catalyst at 650°F to form CO_2 and H_2:

$$CO + H_2O \longrightarrow CO_2 + H_2$$

3. Gas purification. In order to provide a concentrated H_2 stream, the CO_2 is separated from the H_2 by a solvent extraction process.

4. Methanation. Since the presence of any CO or CO_2 in the H_2 stream can mess up some of the applications, a cleanup step converts any traces to CH_4. A catalyst at 800°F is used.

$$CO + 3H_2 \longrightarrow CH_4 + H_2O$$
$$CO_2 + 4H_2 \longrightarrow CH_4 + 2H_2O$$

In some places, a sulfur-free CH_4 (natural gas) feed is not available. In that case refiners substitute heavier hydrocarbons such as propane or naphtha. The equipment design and catalysts are different and the process is less fuel efficient, but it works.

Sulfur Facilities

Hydrotreating creates H_2S streams—a deadly, toxic gas that needs careful disposal. The usual process involves two steps: first, the removal of the H_2S stream from the hydrocarbon streams; second, the conversion of the lethal H_2S to elemental sulfur, a harmless chemical.

H_2S recovery. Until about 1970, most of the H_2S in refineries was used, along with the other light ends, as refinery fuel. The product of burning H_2S is sulfur dioxide (SO_2). SO_2 stinks like rotten eggs, but it doesn't kill anyone. Air quality regulations now limit SO_2 emissions to the extent that almost 100% of the H_2S must be kept out of the fuel systems.

Recovery of the H_2S can be done by a number of different chemical processes. The most widely used is solvent extraction using diethanolamine (DEA) (Figure 15-3). A mixture of DEA and water trickles down a vessel with trays or packing that will distribute the liquid. The gas stream containing the H_2S enters at the bottom. As the streams circulate, the DEA

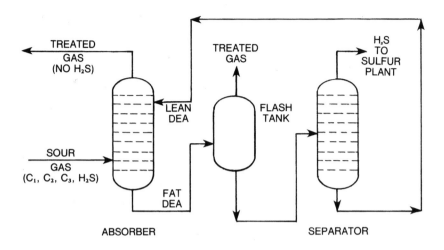

Fig 15-3—Diethanolamine Treater (H$_2$S Recovery)

selectively absorbs the H$_2$S gas. The rich DEA is then fractionated to separate the H$_2$S, which is sent to a sulfur recovery plant. The stripped DEA is recycled. This process is similar to the lean oil/fat oil process for separating C$_2^-$ gas from C$_3^-$ gas described in chapter 7 on gas plants. The difference is that the DEA selectively picks up the H$_2$S, but not any hydrocarbons.

Sulfur recovery. The conversion of H$_2$S to plain sulfur is done in a process first developed by a German named Claus in 1885 (Figure 15-4). There are variations on the process today suited to various H$_2$S/hydrocarbon ratios, but they mostly use a basic two-step, split-stream process.

1. *Combustion.* Part of the H$_2$S stream is burned in a furnace, producing SO$_2$, H$_2$O, and sulfur.

$$2H_2S + 2O_2 \qquad SO_2 + S + 2H_2O$$

SO$_2$ and elemental sulfur, $\vec{S}$, are formed instead of just SO$_2$ because the air (oxygen) admitted to the furnace is limited to one-third the amount needed to make all SO$_2$. All the gases from the furnace are captured so that no SO$_2$ escapes to the atmosphere.

2. *Reaction.* The remainder of the H_2S is mixed with the combustion products and passed over a catalyst. The H_2S reacts with the SO_2 to form sulfur and water.

$$2H_2S + SO_2 \longrightarrow 3S + H_2O$$

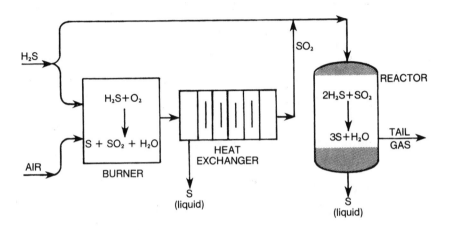

Fig 15-4—Claus Sulfur Plant

The sulfur drops out of the reaction vessel in the molten (melted) form. Most refineries store and ship sulfur in the molten state, although some companies store sulfur by pouring it on the ground (in forms), letting it solidify. Sulfur can be stored indefinitely in this dry state, called a *sulfur pile.*

Claus plants convert about 90% to 93% of the H_2S to S. Local environmental conditions may permit the balance of the H_2S, known as the tail gas (pardon the expression again) to be burned in the refinery fuel system, but not in the US. The tail gas can be processed for high percentage H_2S removal in more elaborate facilities like Sulfreen, Stretford, or SCOT (Shell Claus offgas treating).

Exercises

1. Identify which of these streams are feeds, products, or internal streams in hydrotreating, DEA removal, Claus plants, and an SMR:

H_2S CH_4

S H_2

CO O_2

CO_2 SO_2

16

Isomerization and Dehydrogenation

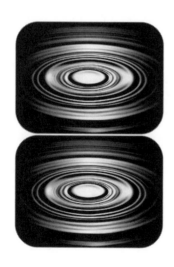

'My dear Bilbo! You are not the same hobbit that you were'.

-The Hobbit, **J.R.R. Tolkien**

I som plants are molecule re-arrangers somewhat like the cat reformer except that they only convert normal paraffins to isoparaffins. Curiously, the C_4 isomerization unit is built for very different reasons than the C_5/C_6 isomerization unit.

A refiner uses a dehydro plant to convert a saturate to an olefin by snipping off one of the hydrogen atoms. Refiners might want to do that if they have plenty of normal or iso-butane but not enough normal or iso-butylene. However, about the only time they can afford a dehydrogenation plant is if they are going to get a big economic boost from the olefin, and that is usually limited to using iso-butylene to make MTBE, a big gasoline octane booster.

Butane Isomerization

A refinery that has an alkylation plant is not likely to have exactly enough iso-butane to match the propylene and butylene feeds. If the refinery has a hydrocracker, there is likely to be surplus iso-butane that is probably blended off to gasoline. Without a hydrocracker, but with a cat cracker and alky plant, a refiner will probably have to supplement the iC_4 supply. The refiner usually has two choices—buy it or make it in a butane isomerization (BI) plant (Figure 16-1).

The process. The feed to the BI plant is normal butane or mixed butanes (iso and normal), which are sometimes called field grade butanes if they come from a gas processing plant. The butanes should not have any trace of olefins that would deactivate the catalyst.

The butanes are fed to a feed preparation column where iso-butane is separated, which is cheaper than pumping it around and taking up space in the BI plant. The high purity normal butane is then mixed with a small amount of H_2 and chloride and charged to a reactor containing a platinum catalyst. The catalyst causes the normal butane to reform itself into its isomer, iso-butane.

The stream coming from the reactor contains about 60% iso-butane, 40% normal butane, and a minor amount of propane and lighter. In a fractionator, the latter are split out and sent to the fuel system; the butanes are recycled to the feed fractionator so that the normal butane can be separated from the iso and re-run.

Yields. When the yields are figured on a net basis, the iso-butane outturn slightly exceeds the normal butane feed due to the volume/weight trick again. In essence, it's normal butane in, iso-butane out, and that's it.

A few plants have been built to convert normal butylene to iso-butylene. Refiners usually choose to put their normal butylenes into their alky plant, but in some unusual situations, they prefer to make so much MTBE that they need all the iso-butylene they can put their hands on and then some. So they convert normal to iso-butylene through isomerization.

C_5/C_6 Isomerization

For a refinery that has problems meeting the octane number of gasoline and has a lot of straight run gasoline around, C_5/C_6 isomerization has appeal. Normal pentane, which has an RON of 62, can be converted to iso-pentane with an RON of 92. Normal hexane goes from an incredibly low 25 RON to about 75; a typical mixture of iso- and normal pentanes and hexanes can be upgraded from 73 to 91 RON.

The process. The boiling points of the four compounds involved, the two normal paraffins and their isomers, are good news and bad news.

iso-pentane	82°F
normal pentane	97°F
iso-hexane	129-142°F
normal hexane	156°F

Iso-pentane boils at the lowest temperature. As in the BI plant, the C_5/C_6 Isom plant shown in Figure 16-2 may have a feed fractionator that separates the iso-pentane from the feed. No sense in carrying along extra baggage that won't be involved in the reaction. At the same time the iso-hexane is often left in the feed. The various iC_6 isomers boil between the normal pentane and the normal hexane and would take two columns or a very tall single column with lots of reflux and reboil to get it out.

A small amount of H_2 and organic chlorides are mixed with the feed,

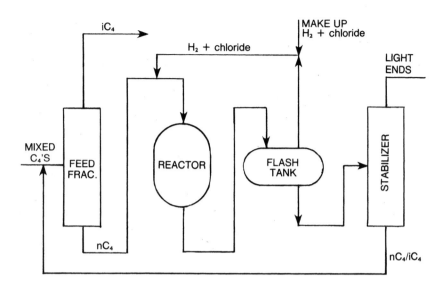

Fig 16-1—Butane Isomerization Plant

which is then charged to a reactor. The catalyst will cause conversion of about half the feed to the corresponding isomers, creating an *isomerate* stream. The reactor product can be fractionated to recycle the normal pentanes to extinction. The C_5's are split from the C_6's and fed back to the feed prep column where the iso-pentane is split out and the normal pentane left in the feed. Again the boiling points of the iso- and normal mix present themselves as a problem. Some refiners put in a column to recycle the normal hexanes along with the normal pentanes. Some just let the normal hexanes stay with the iso-hexane and content themselves to passing the normal hexanes through the process only one time.

Yields. C_5/C_6 isomerization is a little more complex than C_4 isomerization, so about 2-3% light ends, C_4 and lighter, get created in the process. Depending on the extent of recycle, the octane number of the isomerate can be varied from 80 to 91 RON, with the cost of energy (fractionation, pumping) increasing with the octane number. The conversion of the normal paraffins to their isomers on a once-through basis with no recycle runs about 85%.

Dehydrogenation

Refiners sometimes don't have just the right amount of normal and iso-butylene to satisfy their needs. A well-developed trading market exists in many areas for normal butane, and to a lesser extent for iso-butane. If refiners are short of the butylenes, they may want to resort to making them out of butanes in a dehydrogenation unit

For refiners that have an MTBE plant and a voracious appetite for iso-butylene, one way to provide an adequate supply is to take iso-butane and remove two hydrogens. Refiners borrow technology from petro-chemical producers to pick the hydrogen atom off. Once again, a catalyst is the secret to high yields, few side reactions creating unwanted molecules, and low pressures and temperatures to keep down the costs.

Some of the processes for dehydrogenating iso-butane have fixed bed reactors in parallel. The catalyst can be any one of a number of fancy metals like chromic oxide or platinum. The process results in coke deposits on the catalyst, so the reactors are run in alternating cycles. One is in the regeneration phase while the other two or three reactors are in service.

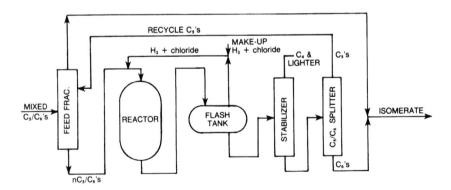

Fig 16-2—C_5/C_6 Isom Plant

In another variation, the one described more completely below, the catalyst moves along with the feed and passes continuously through a regenerator.

The process. For such a simple idea, butane dehydrogenation involves a lot of steps, some of them small, as shown in Figure 16-3. The preparation for the reaction takes as many steps as the reaction itself. A deisobutanizer prepares a concentrated iso-butane stream for the reaction. To assure the catalyst doesn't get poisoned, the stream is passed through a bed that traps all the nitrogen. It then passes through a mole sieve bed to pull out any water that would freeze during the cold parts of the process. The stream is deep chilled and then mixed with a small amount of H_2. The combined stream is then heated and charged to the first reactor.

The reactors are arranged in series, with heaters in between each. The isomerization process is endothermic—it absorbs heat, and in fact the stream cools down as it passes through each reactor and has to be reheated. The platinum catalyst is introduced to the first reactor along with the

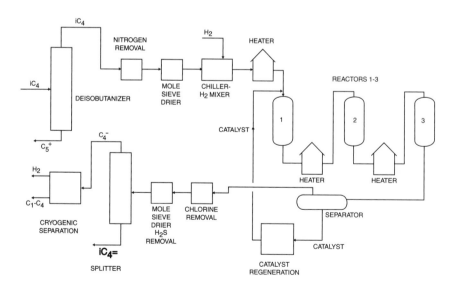

Fig 16-3—**Dehydrogenation of Isobutane**

feed and flows in and out of each reactor. At the end the catalyst is separated from the hydrocarbon and regenerated and recycled to the first reactor. The hydrocarbon, now dehydrogenated, is compressed and treated for H_2O and chloride removal and then for H_2 separation in a cryogenic separation plant.

With all the treating and expensive separation operations, refiners need a big incentive to install the plant to dehydrogenate iso-butylene.

Butadiene. Refiners and petrochemicals companies use a similar dehydrogenation plant to change normal butane and normal butylenes to butadiene, a building block petrochemical used to make synthetic rubber compounds. (See chapter 19 on Ethylene Plants for a little more on butadiene).

Review

Butane isomerization is used to satisfy the feed needs of alkylation by converting normal butane to iso-butane. C_5/C_6 isomerization is a method of increasing the octane number of the light gasoline components, normal pentane and normal hexane, that are found in abundance in light straight run gasoline. Refiners use dehyrogenation to provide themselves with a source of iso-butylene to feed their MTBE plants.

Exercises

1. How are refiners motivated differently in having a BI plant compared to having a C_5/C_6 Isom plant?

2. What has to be added to Figure 16-2 to show recycling of normal hexane?

3. What three ways can a refiner solve the problem of having surplus normal butane but not enough iso-butylene to run an MTBE plant?

chapter seventeen

MTBE and the
Other Ethers

[He] throws aside his paint pots and his words
a foot and a half long.

-Ars Poetica,
Horace, 65-8 BC

I f refiners weren't backed into it starting in the 1970s and later, they probably would never have built any plants to make methyl tertiary butyl ether MTBE—a minor petrochemical generally used at the time as a paint solvent. The "ether" suffix gives MTBE the oxygen atom in the compound, something unnatural to a petroleum refiner who rarely finds oxygen in his molecules.

Refiner interest in MTBE started as governments began to mandate the removal of tetraethyl lead from gasoline blending in the 1970s. Refiners were frantically searching for economical ways to sustain the octane levels of their gasoline pools using environmentally acceptable blendstocks. Fearing that the chemical industry would react sluggishly to the demand for MTBE, refiners began to build their own plants. Then a decade later, governments began to mandate a minimum oxygen content for gasoline to reduce CO and unburned gasoline coming out the tail pipes of cars. That generated further enthusiasm from refiners for MTBE plant construction. To the horror of refiners, some of those governments have since reversed their mandates and now outlawed the use of MTBE in gasoline. They fear contamination of aquifers because MTBE, possibly leaking from underground storage tanks at gas stations, has such an affinity for water. The continuing use of MTBE in gasoline is muddy.

The Feedstocks

The ingredients for MTBE are iso-butylene and methanol. Historically refiners have noticed a perennial surplus of methanol producing capacity around the world. Furthermore, huge surpluses of natural gas in remote locations promise even more methanol supply should the market price ever rise slightly above investment values. In a shrewd move, most refiners have opted to buy their methanol requirements rather than make them.

Iso-butylene supply initially came mainly from the cracked gas streams generated by the cat cracker, plus whatever other units fed gases into the cracked gas plants. The iso-butylene market trades very thinly, so when the cracked gas plant supply is insufficient, a refiner must turn to dehydrogenation to convert iso-butane to iso-butylene, as mentioned in the last chapter.

The Plant and Process

A number of companies offer their own proprietary process designs for making MTBE. Since MTBE came out of the petrochemicals industry, many of them have a more exotic process than refiners normally deal with called *catalytic distillation*. This device combines a distillation column with a reactor in a single vessel. The column has a bed of catalyst in the middle and trays above and below it. The idea is to introduce the feed into the catalyst bed. The catalyst causes a reaction that generates enough heat to cause one reactant to vaporize while the other remains a liquid. The trays of the column, top and bottom, then assure a good clean separation of the two resulting reactants. In the case of the MTBE plant in Figure 17-1, MTBE comes out the bottom of the catalytic distillation column and the unreacted C_4's and methanol go "out the top."

Now for the rest of the vessels and flows in Figure 17-1. The feed consists of iso-butylene, fresh methanol, and recycled methanol. The iso-

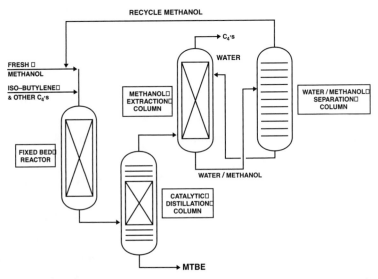

Fig 17-1—MTBE Plant

butylene comes mixed with other C_4's (normal butylenes, iso- and normal butane). The feed passes through the bed of catalyst, indicated by the X in the figure. The catalyst is loosely poured into the vessel to allow easy passage but intimate feed/catalyst contact. The combination of only moderate temperatures and the catalyst start the chemical reaction in Figure 17-2 going between the methanol and the iso-butylene. Almost 90% of the iso-butylene converts to MTBE in this reactor.

The effluent from this fixed bed reactor, both vapor and liquid, goes to the catalytic distillation column where the reaction continues, converting almost all the remaining iso-butylene as the gaseous C_4's and methanol rise through the catalyst. The catalyst in this vessel is loaded in bales, sometimes called "Texas teabags." MTBE, a higher boiling point liquid than the C_4's and methanol, drops out of the bottom of the column as a liquid. The process is run with excess methanol, so the tops include a vapor mixture of the unconverted methanol and the other C_4's.

The easiest way to separate the methanol is to trickle some water through the mixture. Since methanol has an affinity for water but the C_4's have an aversion to it, the H_2O/methanol mixture comes out the bottom

$$CH_3OH \quad + \quad C_4H_3-\overset{\displaystyle CH_3}{\underset{\displaystyle |}{\overset{\displaystyle |}{C}}}=CH_2 \quad \rightarrow \quad CH_3-\overset{\displaystyle CH_3}{\underset{\displaystyle CH_3}{\overset{\displaystyle |}{\underset{\displaystyle |}{C}}}}-O-CH_3$$

Methanol   Iso-butylene   MTBE

Fig 17-2—Chemical Reaction to Make MTBE

of the methanol extraction column and the C_4's out the top. The methanol and water are separated by a simple distillation, with the water recycled back to the methanol extraction column and the methanol recycled back to the beginning of the process.

The Other Ethers

The C_5, C_6, and C_7 olefins that come from the cat cracker have traditionally gone straight into the gasoline blending pool. In some areas the olefin content limits on gasoline content start to bind, or limits on aromatics or benzene content on gasoline, do a job on the octane level of the gasoline pool. In those cases, refiners have taken the C_5, C_6, or C_7 olefins and treated them just like the butylenes and made the corresponding ethers. Sometimes they use ethanol instead of methanol as the co-feed. You can tell the feeds if you look at the names closely:

MTBE	methyl tertiary butyl ether
ETBE	ethyl tertiary butyl ether
TAME	tertiary amyl* methyl ether
TAEE	tertiary amyl* ethyl ether
THxME	tertiary hexyl methyl ether
THxEE	tertiary hexyl ethyl ether
THpME	tertiary heptyl methyl ether
THpEE	tertiary heptyl ethyl ether

* amyl is a prefix for a C_5 olefin

The octane boost from making these iso-olefins into ethers is 10-20 RON, so the economic incentive is not all that bad. In addition, where a refiner needs to put an oxygenate into his gasoline, these ethers qualify. (Each of them has the characteristic, **- O -** , imbedded in the middle of the molecule, just like the MTBE in Figure 17-2.) Mostly though, the environmental constraints from the olefin and aromatics removal are the drivers that make refiners build the expensive facilities to make the higher ethers.

Many of the process designs allow refiners to make any or all the above list of ethers in the same plant. Most of the designs have more or less the configuration as the one in Figure 17-1, but for the higher ethers, one or two additional columns might have to be added for additional separations and recycles.

Exercises

1. What are the feedstocks to produce the eight ethers listed in the chapter?

2. What is the signature characteristic of all ethers?

18

Solvent Recovery of Aromatics

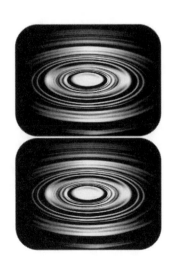

We, only we, are left.

-"Rugby Chapell,"
Matthew Arnold

I n 1907 a man named Edeleanu developed a process by which aromatic compounds could be preferentially removed from a hydrocarbon mixture. The process worked even in cases where the boiling points of the aromatic compounds were the same as the other compounds in the mixture. Distillation under those conditions wouldn't work, so Edeleanu's innovation made a major breakthrough.

Applications

Removal of aromatic compounds can be desirable for two different reasons. Either the aromatics have detrimental effects on the quality of the mixture they're in, or the aromatics are worth more if they're separated than if they're not. You have already seen a couple of examples:

- Gasoline now has a max benzene spec
- Aromatic compounds in kerosene cause unacceptable smoke points

Other more specialized applications include:

- Kerosene range solvents that are aromatics-free or aromatics-laden have various industrial applications
- Separated benzene, xylene, and toluene have numerous chemical applications
- Removing aromatics from heavy gas oil stocks can improve the lubricating oil characteristics

Processes

The solvent recovery process is based on the ability of certain compounds to dissolve certain classes of other compounds selectively. In this case, certain solvents will dissolve aromatics but not paraffins, olefins, or naphthenes. The reasons the process works are a story you won't read here.

The first co-requisite that makes this approach successful is that the solvent with the extracted compounds dissolved in it readily separates itself from the starting hydrocarbon mixture. The second co-requisite is that the solvent and the dissolved extract can easily be split in a fractionator.

Take kerosene as an example, one that has a lot of aromatic compounds in it. To half a beaker of kerosene add half a beaker of a solvent—in this case, liquid SO_2. After mixing, the liquid will separate into two phases with the kerosene on the bottom and the SO_2 on top. The kerosene on the bottom will fill less than half the beaker. The SO_2, because the aromatic compounds have dissolved in it, will take up more than half the beaker.

If the SO_2 is poured off, the aromatic compounds can be "sprung" from it by simple distillation. This two-step process is *batch solvent processing*.

Knowing how a simple batch process works, a continuous flow process is easy to conceptualize. In Figure 18-1, a three column system is shown. The feed is introduced as a vapor into the lower part of a vessel or column with a labyrinth of mixers inside. (Sometimes the mixers are

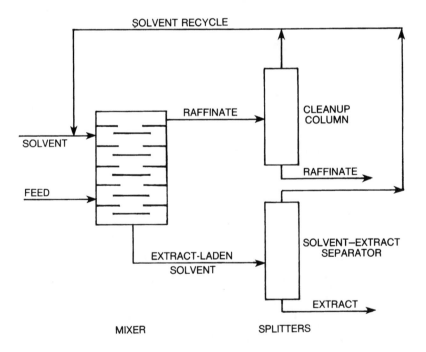

***Fig 18-1*—Solvent Recovery Process**

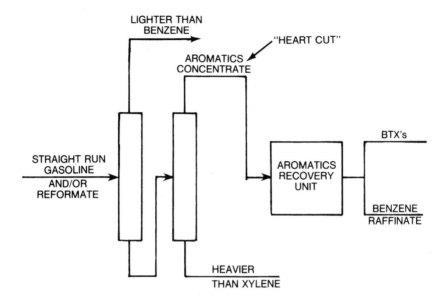

Fig 18-2—BTX Recovery

mechanically moved, such as in a *rotating disc contactor*). The solvent is introduced as a liquid near the top. The solvent works its way towards the bottom of the vessel, dissolving the *extract* as it goes along. The rest of the hydrocarbon, which rises to the top, is called *raffinate*.

Two columns handle the streams coming out of the mixer. One column cleans up any small amounts of solvent that may have followed along with the raffinate. The other column separates the solvent and the extract. The solvent from both columns is recycled to the top of the mixer.

Some of the solvents used in various applications are listed below:

- Kerosene treating: liquid SO_2, Furfural®

- Lubricating oil treating: liquid SO_2, Furfural®, phenol, propane (separates paraffins from asphaltenes)
- Gasoline: Sulfolane®, phenol, acetonitrile, liquid SO_2

Benzene and Aromatics Recovery

The most widespread application of solvent extraction is used in BTX recovery, especially for benzene. To make the process efficient, the feed to the process is pared down to an *aromatics concentrate* by making a *heart cut* from a reformate or straight run gasoline stream as shown in Figure 18-2. The aromatics concentrate then has a large benzene content, making the extraction process more efficient.

One bit of nomenclature is often misleading. *Benzene raffinate* contains no benzene. It is the leftover of the aromatics concentrate after the goodies (benzene) are removed. As a result, benzene raffinate becomes an acceptable gasoline blending component and the separated benzene becomes a chemical feedstock.

Exercises

1. Which has a higher concentration of benzene in it as it leaves the solvent extraction vessel, the benzene raffinate or the solvent?

2. What are the three essential features of a good solvent for extraction?

nineteen

Ethylene Plants

Soone hot, Soone cold, nothing violent is permanent.

-"Petite Pallace,"
George Pettie

T he closest refiners come to becoming chemical companies happens when they build ethylene plants. That's the reason why integrated oil/chemical companies built so many large ethylene plants-they bridged the gap between the two. Ethylene plants are better called *olefins plants*, but are variously referred to as ethylene crackers (a misnomer), steam crackers (another misnomer, although steam *is* mixed with the feed) or something-crackers, with the suffix denoting the feed, as in ethane cracker.

Olefin plants can be designed to crack a number of feedstocks. They usually fall into the following categories:

ethane
ethane/propane mix
propane
butane
naphtha
gas oil

The original olefin plants were built to produce ethylene to supply the growing appetite of the chemical industry for this basic building block. The propylene coming from an olefin plant attracted less interest. Propylene in abundant amounts was always readily available to be filched from refinery alkylation plants. Many early US olefins plants were designed to crack ethane, or ethane and propane, first because ethane was abundantly available from natural gas and second, because of the high yield of ethylene from ethane, as you can see in Table 19-1.

| | Pounds of Product per pound of feed | | | | |
	Ethane	Propane	Butane	Naphtha	Gas Oil
Ethylene	0.77	0.40	0.36	0.23	0.18
Propylene	0.01	0.18	0.20	0.13	0.14
Butylene	0.01	0.02	0.05	0.15	0.06
Butadiene	0.01	0.01	0.03	0.04	0.04
Fuel Gas	0.20	0.38	0.31	0.26	0.18
Gasoline	-	0.01	0.05	0.18	0.18
Gas Oil	-	-	-	0.01	0.12
Fuel Oil	-	-	-	-	0.10

Table 19-1—Ethylene Plant Yields

In other parts of the world naphtha emerged as the preferred flavor of feed. The demand for gasoline and therefore the naphtha cuts has always grown much slower outside the US. In the 1970s, olefin plants were built to crack gas oil, producing as by-products some high octane gasoline blending components. A number of huge olefin plants, as large as medium-sized refineries, now are integrated into refineries and produce a significant amount of the gasoline.

Process

Ethane/propane crackers, as shown in Figure 19-1, are the simplest design, but they demonstrate the fundamentals. The ethane and propane can be fed separately or as a mixture to the cracking furnaces where the short residence time and high temperature yield a high volume of ethylene. The quench downstream of the cracking furnaces stops the cracking

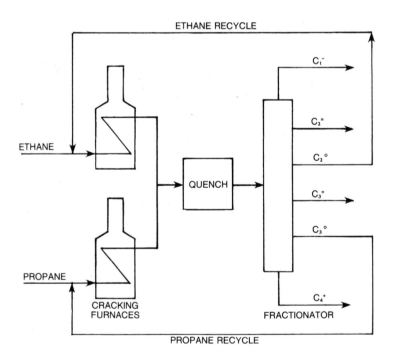

Fig 19-1—Olefins Plant: Ethane-Propane Cracker

process from taking the feeds all the way to methane. But for the operating conditions and feeds, and a few bells and whistles, olefin plants are just plain old thermal crackers.

In one pass through the cracking furnace, not all the ethane and propane disappear. Downstream in the product fractionator, the ethane and propane are split out and recycled to the feed. Generally, the ethane is recycled to extinction, but some of the propane goes with the propylene.

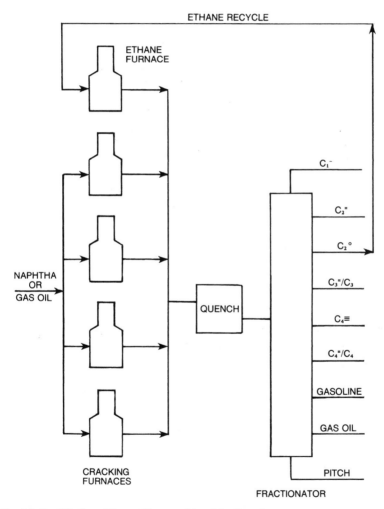

Fig 19-2—Olefins Plant: Heavy Liquids Cracker

The olefin plants designed to crack heavier feedstocks need extra separation facilities. Table 19-1 shows that ethylene and propylene make up only about 35% of the yield from cracking naphtha and gas oil. The rest has to be separated in a series of columns and a cryogenic unit to recover the ethylene. Figure 19-2 simplifies the scheme by showing only a single fractionator. The yield in Table 19-1 assumes that the ethane is recycled to an ethane furnace.

Refinery Interactions

Olefin plants sometimes provide a home for a number of junk streams in a refinery. For example, the dry gas stream from a cat cracker is usually sent to the fuel system, even though it contains ethane, ethylene, some propane, and propylene. These components can be recovered by commingling them with the furnace outturns of the olefin plant before it heads for the separation section.

As another example, some of the naphthas in the gasoline pool are of such low octane they get shut out of the gasoline pool as a blending component. Benzene raffinate is an example. These streams can make attractive feedstocks for refiner-owned olefins plants, turning trash into treasures.

On the outturn side of the olefin plant, refiners find some other complementary opportunities. Beside the gasoline range products, the butanes and butylenes can generally find a home in the refinery processes.

Products. Ethylene is normally sold as a 99$^+$% purity stream. Deliveries are almost always by pipeline. Propylene is bought and sold in three different commercial grades that differ by the accompanying amount of propane mixed in. Polymer grade propylene is 97-99% propylene; chemical grade is 92-95%; refinery grade is 50-65%. Refinery grade propylene is the natural propane/propylene combination coming from cat crackers, the biggest source of propylene in a refinery. Chemical grade propylene comes out of naphtha and gas oil at about 92%, with 8% propane. Either of these two grades has to be fractionated to get the higher purity polymer grade propylene.

A chemical stream that turns up here, but not in the refinery sector, is butadiene (Figure 19-3). This chemical is in the *di-olefin* family and has two double bonds with the formula C_4H_6. The two double bonds make it particularly reactive, especially for creating plastics and rubber compounds.

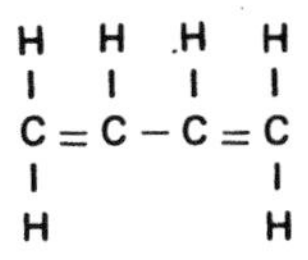

$$
\begin{array}{cccc}
\text{H} & \text{H} & \text{H} & \text{H} \\
| & | & | & | \\
\text{C} = \text{C} & - & \text{C} = \text{C} \\
| & & & | \\
\text{H} & & & \text{H}
\end{array}
$$

Butadiene

Fig 19-3—Butadiene

Exercises

1. How much feed, in barrels per day do you need to run a typical worldscale olefins plant producing a billion lbs per year of ethylene from the following feeds:

    ethane (3.2 lb/gal)
    propane (4.24 lb/gal)
    naphtha (6.4 lb/gal)
    gas oil (7.3 lb/gal)

2. A company has an ethane/propane cracker with a capacity of 500 MM lbs per year of ethylene. Presently it is cracking a mixture of 70% (by volume) ethane and 30% propane, running the plant at capacity. Suddenly the propylene market turns sour, and the company wishes to produce only 20 MM lbs per year of propylene. How much feed were they cracking, how much propylene were they producing, and how much ethane should they substitute for propane?

20

Simple & Complex Refineries

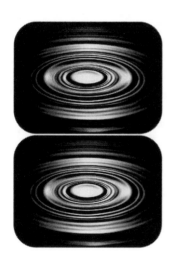

Aero roussimo nostro, Simplicitas.
(Simplicity, most rage in our age).

-"Ars Amatoria,"
Publius Ovidus Naso

A look at the economics of simple and complex refineries offers one way to tie the essential parts of this book together. After all is said and done, refineries are about making money, and some refineries make more than others, just because of the assembly of processing units in them. This chapter discusses why.

Measuring Profitability

In the early 1980s the businessmen running refineries made an intellectual breakthrough that coincided with the revolution underway in information technology. They recognized that the prices for crude oil, gasoline, distillates, and residual fuels connected to each other in a profound way. The kinds of hardware in refineries around the world not only had impact on the profitability of refineries. They set, in large part, the margins between crude and products and the differences between crudes themselves.

The constellation of relationships tying crude to products had a prerequisite. Prices had to be free to respond to market forces—the changing preferences of consumers, the sometimes-capricious wishes of crude producers, and the influence of other fuels, especially natural gas and coal. The rise in spot markets during the 1970s provided the necessary setting.

The essence of the structure around crude and products pricing and profitability was the relative ability of a refinery to produce various slates of products. The key concept was simple—*the more valuable products a refinery makes, the more it can pay for crude oil.* From this theorem refiners adapted an old set of nomenclature—simple, complex, and very complex refineries.

Definition:
Simple refinery—crude distillation, cat reforming, and distillates hydrotreating
Complex refinery—simple refinery plus a cat cracker, alky plant, gas processing
Very complex refinery—complex refinery plus a coker which eliminates residual fuel production

Origins of "Complexity"

In the 1960s, when refiners were still building refineries as fast as they could, Wilbur Nelson devised a shorthand for characterizing how much plant and equipment a refinery had. He measured the cost and throughput of every refining unit (cat cracking, reforming, alkylation, etc.) relative to distilling. Then he calculated an index that captured both relative costs and throughput and called it the complexity factor.

For example, a simple refinery with a distilling column, a cat reformer to upgrade the heavy naphtha, plus some treating of gas and distillates might have a complexity factor of 2.1. Distilling would make up 1.0; the cat reformer had only 15% as much throughput, but cat reforming capacity per barrel costs 4 times as much as distilling. That adds .15 X 4.0 = 0.6 to the complexity factor. The treating might add another 0.6 for a total of 2.1.

A more elaborate refinery that also has a cat cracker, alky plant, maybe a thermal cracker, a gas plant, and hydrotreating might have a complexity factor of 9 or 10. Nelson thought of refineries with a complexity factor of 2-5, variations of the so-called hydroskimming refinery shown in Figure 20-1, as simple refineries. Complex refineries, as in Figure 20-2, had complexity factors in the 8-12 range. Refineries that had cokers or ethylene plants pushed their complexity factors up to 15 or more, very complex

In recent years, active buying and selling refineries has breathed new life into the Nelson Index. Describing the market for refineries by measuring the dollars per barrel of distilling capacity doesn't adequately describe the value involved. Many analysts now measure the ongoing transactions in dollars per complexity-barrel, which takes into account both capacity and the replacement cost of all the processing units in the refinery.

$$\text{Dollars per complexity barrel} = \frac{\text{Transaction value}}{\text{Distilling capacity x Complexity}}$$

Nelson's intent was to provide a scheme to compare various

investment options. Keep in mind, the name of the game in the 1960s was "keeping up with demand." Even though his idea dealt with only the investment side of the equation, it set out the framework for later analysis of operating costs, revenues, and profitability.

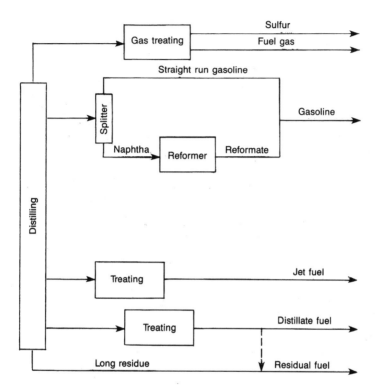

Fig 20-1—Hydroskimming (or Simple) Refinery

The simple refinery looks more or less like Figure 20-1. The addition of a cat cracker in Figure 20-2 re-defines it as a complex refinery. (The cat cracker dominates that refinery even though it has a visbreaker and an isom plant. On the margin it will behave economically like a complex refinery). To get to a very complex refinery, you need only add a coker to Figure 20-2.

As you move from simple to complex to very complex, the gasoline yield (and gain) goes up and the resid yield goes down. Table 20-1 shows the typical yields from running a medium sulfur and medium weight crude like West Texas Sour or Arab Light.

	Simple	Complex	Very Complex
Gasoline	30	50	60
Jet Fuel	10	10	10
Distillate Fuel	20	25	25
Residual Fuel	35	10	—
LPG	—	3	4
Coke	—	—	3
Refinery Fuel	8	12	13
Gain	(3)	(10)	(15)

Table 20-1—**% Refinery Yields from Medium Crude**

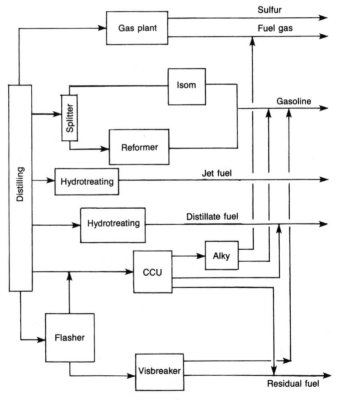

Fig 20-2—**Gasoline (or Complex) Refinery**

To begin to understand how refinery configuration can influence the relationships between crude prices and between crude and product prices, start with the calculation in Table 20-2. The sum of the product yields times the refinery gate prices gives the total revenue. Subtract the operating cost and the crude cost and you get the refining margin.

	Simple refinery			Complex Refinery		
	% Volume	$/Barrel	$/Barrel	% Volume	$/Barrel	$/Barrel
Gasoline	30	22	6.60	50	22	11.00
Jet Fuel	10	18	1.80	10	18	1.80
Distillate Fuel	20	17	3.40	25	17	4.25
Residual Fuel	35	11	3.85	10	11	1.10
Refinery fuel (Gain)	8 (3)	—	— — —	12 (6)	— —	— —
Total Outturn /Revenue	100		15.65	100		18.15
Medium Crude Cost	100	15	15.00	100	15	15.00
Operating Cost		1	1.00		2.50	2.50
Refining Margin			(0.35)			0.65

Table 20-2—Refining $15 Medium Crude

You can note the following things about processing the same type of crude in two different refineries:

- The light oil yield is lower in the simple refinery
- Lower revenue comes with lower light oil yields
- Operating costs are lower in a simple refinery
- Profitability of the simple refinery is lower *(negative, as a matter of fact)*

The extra margin, of course, is what the owner of the complex refinery hopes for and expects as compensation for spending all that money for additional processing units. It doesn't always turn out that way, as you will note in a few pages.

Now, move on to Table 20-3 where the same two refineries are now processing a very heavy crude, like Mayan from Mexico or Arab Heavy, instead of the medium crude. You can make the following observations about this example:

- Compared to running the medium crude, the light oil yield goes down big time in the simple refinery
- Even the complex refinery does not have enough residue reduction capacity to keep the residual fuel yield down and the light oil yield up
- The heavy crude price is lower than the medium crude price, low enough to make the complex refinery more profitable
- The simple refinery simply does not have enough conversion capacity to take advantage of the difference between the light oil prices and the residual fuel price (*the light/heavy differential*)

	Simple refinery			Complex refinery		
	% Volume	$/B	$/B	% Volume	$/B	$/B
Gasoline	10	22	2.20	50	22	11.00
Jet Fuel	5	18	0.90	5	18	0.90
Distillate Fuel	17	17	2.89	30	17	5.10
Residual Fuel	63	11	6.93	10	11	1.10
Refinery Fuel	10	--	--	15	--	--
(Gain)	(5)	--	--	(10)	--	--
Total Outturn/ Revenue	100		9.84	100		18.10
Heavy Crude Cost	100	12	12.00	100	12	12.00
Operating Cost		1	1.00		2.50	2.50
Refining Margin			(3.16)			3.60

Table 20-3—Refining $12 Heavy Crude

Now think of a hypothetical world with only these two refineries, running only these two crudes. The market forces make the world unstable. One or more of the following has to happen:

- The simple refineries go out of business
- The crude producers lower their prices so the simple refiners stay in business. Or maybe they raise their prices to capture some of the margins the complex refiners are enjoying. Or maybe product prices go up and simple refiners stay in business. That all depends on whether the demand for products needs the least efficient, simple refinery capacity to keep running
- The medium crude producers lower their price to become more competitive and get more market share from the heavy crude producers
- The price of residual fuel increases as simple refiners go out of business and dry up resid supply
- The light oil prices come down as complex refiners run more crude, producing even more light oils

Any or all of the above changes, and more, put continuous pressure on the crude and products market and cause both crude and product prices to *change relative to each other.*

Tracking Profitability

The world isn't composed of two refineries and two crudes. The numbers are more like thousands. To bring some order to the chaos, refiners try to follow the relationships setting the prices of crude and products. For example, they may want to look at how competing refiners are fairing. They model the simple, complex and very complex refineries and calculate the results of running a selection of light, medium and heavy crudes. The results might look like Table 20-4, three sets of crudes in all three refineries in one market. The calculation uses the same set of product prices but the yields are unique to each refinery-crude combination.

U.S. Gulf Coast Refining Margins

	Crude Price $/barrel	Refining Margins – $/barrel		
		Simple	Complex	Very Complex
Light Crudes				
Louisiana Sweet	18.00	0.10	0.30	0.35
Nigerian Light	18.50	(0.20)	0.10	0.40
Medium Crudes				
West Texas Sour	17.00	(0.35)	0.65	1.52
Arab Light	17.50	(0.95)	0.45	1.10
Isthmus (Mexico)	17.25	(0.50)	0.40	1.20
Heavy Crudes				
Mayan	12.00	(3.16)	3.60	4.60
Arab Heavy	12.50	(3.33)	3.40	4.00
Venezuelan	12.25	(3.49)	3.20	4.25

Table 20-4—U.S. Gulf Coast Refining Margins

What does this table, which reflects a hypothetical point in time, say about prices and margins?

- Simple refineries break even, more or less, running light, sweet crude, but not the heavy, more sour crudes. They make too much high S resid
- The very complex refineries make a lot of money running the heavy crudes. The heavy crudes may be underpriced because they are in over-supply. Even the complex refineries make an adequate margin running heavy crude
- The medium crudes are profitable in complex refineries but they may have a tough time competing for space in the very complex refineries. Those will probably fill up on heavy crudes
- The light crudes may be too expensive and might have to come down in price to compete with heavier crudes

While Table 20-4 gives an overview of how all crudes are competing with each other, crude producers probably get nervous looking at all these "mights" and "may be's." For that reason they track their own *net-backs*, the attractiveness of their own crudes. Netbacks measure the return producers receive from product prices after refiners take out their oper-ating costs and transportation expenses. The calculation takes into account the location of the market (the refinery site), the refinery margin calculation, and the transportation from the wellhead to the refinery. Table 20-5 shows a netback calculation for medium crude refined in the US Gulf Coast on a specific day.

$/Barrel

	Simple	Complex	Very Complex
Refinery Revenues Less:	15.65	18.15	19.52
Operating Cost	1.00	2.50	3.00
Transportation	0.30	0.30	0.30
Netback to the wellhead	14.35	15.35	16.22
Compared to the Medium Crude price	15.00	15.00	15.00

Table 20-5—Netback Calculations

As before, the refinery revenues are modeled numbers, because no one knows exactly what the price setting or marginal refinery looks like. It is a best guess based on the equipment in various refineries, gathered from the *Oil & Gas Journal* and other sources. From this snapshot in time, producers can see how the market perceives the value of their own crude, especially in relation to the current crude price. (You'll note the differences in netbacks have to reflect the same differences in the refinery margins).

Of course, every day or week or month the revenues change as the product prices change, which gives rise to changes in the netbacks. (The transportation expense could vary too, especially if it reflects a long sea haul half way around the world.) To keep tabs on the ebbs and flows in the marketplace, producers will want to plot something like the chart in

Figure 20-3. If that were a real life example for medium crude producers in an area, they might conclude several things:

- They had chosen to continuously price themselves out of the simple

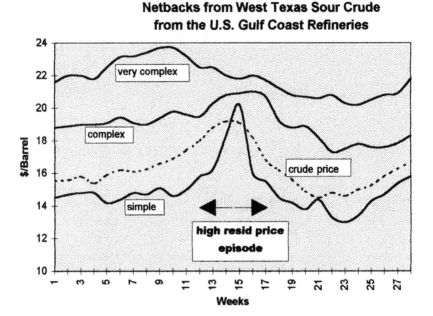

Netbacks from West Texas Sour Crude from the U.S. Gulf Coast Refineries

Fig 20-3—Netbacks

refinery market segment, except for one short period. In week 15, *e.g.*, they might have seen residual fuel prices jump up so much that running crude just to make resid was attractive. Simple refineries with big resid yields were almost as profitable as very complex refineries that make none

- Netbacks are creeping up again at the end of the period. But this time, as another example, they observe that light oil products are increasing faster than resid prices. That's why very complex refinery netbacks are increasing faster than crude price and than the netbacks from complex and simple refineries

Every segment of this chart has a story to tell producers. Similarly, the mirror image of this chart would be valuable to refiners. It would show for, specific refineries the margins from various crudes over time. That would help those refiners understand and anticipate the best crudes to line up for their own operations.

Same Refinery-Different Modes

Why even think about simple refineries when they seem to be unprofitable anachronisms? In almost every example they seem to lose money. The facts are there are not very many simple refineries in the world and hardly any in the developed countries where gasoline is a big part of the market. Product yields and crude composition demand conversion capacity be included in refinery configurations. However a closer look at how refiners put together the refining jigsaw reveals a subtle observation—*at high utilization rates, almost every refinery operates like a worse refinery than it does on the average.*

The units downstream of distilling all have fixed capacities. They fill up as a refiner increases crude runs. The cat cracker and coker, the criti-

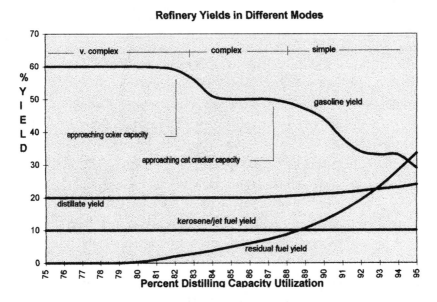

Fig 20-4—Refinery yields in different modes

cal conversion units in complex and very complex refineries, might be matched to the distilling capacity. But then again, the amount of cat feed for the cracker and pitch for the coker depends on what type of crude a refiner feeds to the distilling unit. The heavier the feed, the more quickly those two units fill up.

On top of the other breakthroughs about refinery economics came the realization that any complex or very complex refinery might itself run in two or three modes from time to time. Figure 20-4 illustrates how that might come about.

As a refiner runs the distilling unit at ever-higher rates, typically the coker fills up. When that happens, the next increment of crude does not return the same profit as the previous volume. The refinery slips from a very complex mode to a complex mode.

As the distilling unit processes more crude and the cat cracker fills up, the returns slip again. The last increment of crude run through the distilling unit gets simple yields and lower margins. The same refinery can operate on the margin in any of the three modes. When a refiner looks to buy that last cargo of crude, the capacity utilization dictates the incentive.

What Sets Prices

By this time you are probably exasperated and mumbling, "So what *does* set the product and crude prices?" You have seen how different types of refineries simultaneously enjoy different margins (Table 20-2.) You have seen how the refining margins change in relationship to each other over time (Fig. 20-3). Now you can extrapolate Figure 20-3 to the whole industry to understand what sets prices.

Industry product demands and crude supply fill up industry's capacity just as refiners fill up their own plants. If product demands increase, the most profitable (very complex) refineries fill up first, then the complex refineries and the simple. If product demands are high enough, the refining margins have to increase to keep the simple refineries in business. In Table 20-4, that's not happening. It looks like the very complex refineries have filled up and the marginal refineries are the complex ones.

All types of crudes give a reasonably positive margin in the complex refinery. At the same time, the very complex refineries are making a relatively large bundle of money because they have a coker that turns resid into gasoline. They just keep on refining crude and making money. But they don't, in this case, set the margins. Their cokers are full and if they haven't filled their cat crackers, they are undoubtedly heading that way.

What does Table 20-4 say about setting *crude* prices? The complex and very complex refineries make big margins running heavy crudes. That suggests there's plenty of heavy crude around. In contrast, they make puny margins on light crudes, which suggests light crudes are in short supply and/or are over-priced. Medium crudes, making medium margins in complex and very complex refineries, are the marginal crudes and will put pressure on the light crude producers to increase production and/or reduce their prices.

Suppose something changes—heavy crude production drops. Heavy crude prices would rise in reaction. Resid from complex and simple refineries would decline as medium or light crudes replace heavy crude. Resid prices would rise. Cokers and very complex refineries would no longer look so attractive. They may be full but their margins would decline as heavy crude and resid prices would go up (and light oil prices decline, keeping medium crude/complex refinery margins the same.) Margins for medium and heavy crudes in complex and very complex refineries would converge. If the heavy crude shortage becomes severe enough, the complex margins would rise above very complex refineries. Together with heavy crudes, they would become the combined price setters.

All that requires you keep a multitude of moving parts in mind simultaneously—not easy to do without careful thought. As the dynamics of the oil market change the relationships every day, the frantic business of crude selection requires continuous, rigorous analysis of industry margins to get the best combination.

All the while the general level of crude oil prices moves up and down, virtually independently of all these margin changes. But that's another story.

Review

A refiner's margins are a function of his own refinery's processing units, the crudes he buys and the prices he receives for his products. But the prices and costs are set by an elaborate constellation of industry refinery economics that mimic his own. The relationship between product prices and crude prices has more moving parts than the refineries themselves. And that makes complexity watching vital as refiners try to catch early warning signals for significant profitability changes.

Exercises

1. How can the netback from a simple refinery ever be higher than the netback from a complex refinery?

2. Why do some people say the light/heavy differential is the price of a coker?

3. If light oil prices are going down faster than crude oil prices are going down and resid prices are staying the same, what happens to the margins in simple, complex, and very complex refineries?

Crude Oil, Condensate, and Natural Gas Liquids

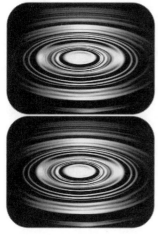

*Get your facts first, and then you can
distort them as much as you please.*

-Mark Twain

T he raw materials coming into a refinery from the oil patch have some names that are colorful, but sometimes not too descriptive. Some sound a lot alike. What's the difference between natural gasoline, natural gas liquids, liquefied petroleum gas, liquefied natural gas and compressed natural gas; between crude and condensate?

Oil Patch Operations

The simplistic cross-cut of the oil patch in Figure 21-1 shows what goes on in a producing oil field. The underground accumulations of hydrocarbon can be in several forms. The well on the left has tapped an oil reservoir that has a *gas cap* at the top. If the well is being produced from just the gas zone, the production is called *gas well gas*. Similarly, but not shown, if a well taps into a reservoir with natural gas only, it produces gas well gas. This hydrocarbon mixture is predominantly methane, but 15-20% of the material could be heavier hydrocarbons, ranging up to the gas oils. Sometimes the production has almost no hydrocarbons heavier than butane, in which case it is referred to as *dry gas*.

As the gas comes out of the well and moves down a pipeline a short way, it cools down and the heavier hydrocarbons liquefy. In the example, the mixture goes into a vessel called a *field separator*, sometimes referred

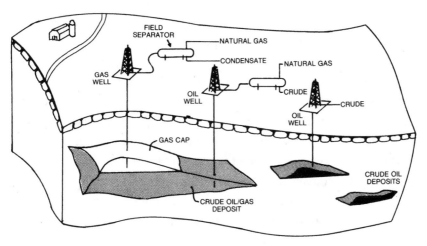

Fig 21-1—Oil Patch Operations

to as a "wide spot in the line." As the vapor and liquid enter the larger space the "beer bottle" effect happens. The pressure drops, and whatever light ends are dissolved in the heavier liquids vaporize and bubble out, just like the carbonation in a beer when you pop the top. Natural gas is drawn off the top of the separator. The bottoms are *condensate*, liquids in the C_4 to gas oil range.

The well in the middle has tapped the same formation, but the production is crude oil with some dissolved natural gas. At the surface, another field separator will split this mixture into natural gas and crude oil. Since the natural gas comes along with the crude oil, it is referred to as *associated gas*. The production from the gas cap or a gas/condensate well would be *non-associated gas*. The distinction is important not chemically, but from a management point of view. Natural gas consumption varies with the seasons. Often producers cannot find enough storage to produce their natural gas at maximum rates all year. The penalty to shut-in associated gas is huge because of the loss of production of the crude oil, compared to the small amount of condensate lost, if any, from non-associated gas.

The well on the right has tapped another formation in which there is only crude oil, no gas. No separator facilities are necessary.

Further processing. Crude oil and condensate can be collected and transported by truck, tank car, or pipeline. Natural gas almost always moves by pipeline. If there's no pipeline around, associated natural gas might have to be reinjected back into the ground—an expensive proposition—to keep the associated crude oil production coming. The practice of continuous *flaring* associated gas to maintain crude production, so prevalent in the Middle East and West Africa, is illegal in the US.

When crude oil is moved by pipeline, often a number of different grades of crude oil from different formations are brought together in a single mix. If it is convenient, condensate is mixed in with the crude as well.

Despite the fact that the natural gas has gone through a field separator, some hydrocarbons heavier than the methane-ethane range may still remain. The natural gas may be processed in a gas processing plant for the recovery of the *natural gas liquids* (NGLs). The NGLs consist of ethane, propane, butanes, and *natural gasoline,* which is a mixture of pentanes through C_9's or C_{10}'s. Sometimes the natural gasoline and butane content could be large enough that during the cold winter months they may condense (liquefy) in the natural gas transmission lines. The pipeline pumps could be made inoperative, shutting down the pipeline. To prevent this, very "wet" natural gas *must* be processed in a gas plant to make it transportable and marketable by removing the heavier NGLs, usually all the C_5's and heavier and some of the C_4's. (Did you remember that normal butane liquefies at about 31°F?)

Gas Processing Plants

Four basic schemes can be used to recover the NGLs:

- refrigeration plant
- cryogenic refrigeration
- lean oil absorption
- dry bed adsorption

In the *refrigeration plant,* the liquids-laden gas is cooled to a temperature somewhere in the range of 15°F down as low as -40°F. At the lower temperature, about 70% of the ethane, 90% of the propane, and all of the butanes and heavier liquefy and can be separated for fractionation.

Ethane makes an attractive feedstock for ethylene plants and producers sometimes want to extract as much ethane as they can to sell into this market. *Deep-cut* ethane recovery operation is done in *cryogenic* refrigeration plants. In this operation, an apparatus called a turbo-expander is used to drop the temperature of the liquids-laden gas to -150°F to -225°F. Under these conditions, 90-95% of the ethane and all the propane and heavier drops out of the gas.

The older gas processing plants are the *lean oil absorption plants,* of the same design as the refinery gas plants. A plain lean oil absorption plant will recover about 70% of the propane and all the butanes and heavier. If refrigerated lean oil is used, all of the propane can be recovered plus 50-75% of the ethane.

Dry bed adsorption is an interesting process that is used when a gas is to be processed only for dew point control (removing only the heavier liquids that might condense in transit). Many gas sales contracts require that the dew point (the temperature at which droplets form) be no higher than 15°F at the pipeline pressure of 800 psi. That will require all the natural gasoline and some of the butane to be removed.

Certain porous materials like activated charcoal, silica gel, and alumina have the nice property of causing liquids to condense on their surfaces. The process is called adsorption since the liquid ends up on the surface. That contrasts with *absorption* where the liquids end up inside the absorbent. In the adsorption process, after a sufficient amount of liquids (in this case NGLs) are condensed on the adsorbent from the gas, the flow is diverted to another vessel of adsorbent. The adsorbed materials in the first vessel are then driven off the adsorbent with steam, collected, and condensed. The adsorbent is then ready for more adsorbing.

Dry bed adsorption will recover 10-15% of the butanes and 50-90% of the natural gasoline. By the way, it's adsorption and adsorbent, just like it's absorption and absorbent.

Transport and Disposition

The market for the oil production coming out of the oil patch production is refineries. As mentioned above, most of it moves in pipelines or tankers to the refinery centers. In the US, many smaller wells produce not in hundreds and thousands of barrels a day but in tens of barrels. Oil begins its move to market by truck from these wells.

Crude oil can absorb virtually all the condensate anyone wants to put in, as a convenient and efficient way to transport. In addition, most crude oil pipelines can take 5-15% butane and natural gasoline from gas plants as well. Crude oil with butane and/or natural gasoline injected in it is called *volatiles-laden crude*. The vapor pressure limits on the pipeline or tanker determine the maximum amount of butane that can be injected in the crude.

Propane and ethane are often handled differently than butane and natural gasoline. Since propane is a marketable product, it is often separated at the gas plant and moved away by truck, tank car, or pipeline. Ethane, on the other hand, is pretty much limited to pipeline movement because of its vapor pressure, density, etc. Sometimes it is pipelined, segregated, sometimes mixed with the other NGLs. In that case the mixture of natural gasoline and lighter is separated from the gas in the gas plant, but is not further fractionated. It moves as a mixture called *raw make, raw mix, or EP mix*. Eventually, these will end up being fractionated at a plant site closer to the market or a more convenient redistribution point.

Constituents of Various Oil Patch Streams

	C_1	C_2	C_3	C_4	$C_5 - C_{10}$	$C_{11}{}^+$
Crude Oil				X	X	X
Condensate				X	X	X
Natural Gasoline				X	X	
Natural Gas Liquids		X	X	X	X	
EP Mix		X	X			
Raw Make or Mix		X	X	X	X	
LPG			X	X		
LNG	X					
CNG	X					
Natural gas	X	X				

Table 21-1—Constituents of Oil Patch Production

Liquefied and Compressed Natural Gas

When producers have no access to a pipeline to transport their natural gas to market, such as Nigeria or Indonesia, they can resort to liquefying the gas and transporting it by ship as liquefied natural gas (*LNG*). LNG plants are very capital intensive and require huge volumes of natural gas reserves to justify them.

Some small amounts of natural gas are liquefied by a few natural gas utilities as a way of storing them to be used during peak demands for natural gas during the winter.

Compressed natural gas (CNG) is methane stored or transported as a gas under high pressure. *CNG* has a small, growing use as transport fuel, particularly in city buses. It has virtually no sulfur in it and generates no noxious fumes to spew out the tail pipe and so wins the favor of the environmentally concerned.

chapter twenty-two

Fuel Values—
Heating Values

If you can't stand the heat, get out of the kitchen.

-Harry S. Truman

M any of the economic forces on the various petroleum products are tied to the amount of heat the individual products yield when they are burned. Indeed, the choice of which streams to use as refinery fuel must take into account the market value of the streams and the heating value.

Thermal Content

When a hydrocarbon is burned, two things happen: a chemical reaction takes place and heat is generated. Typically, the chemical reaction is a transformation of the hydrocarbon and oxygen into H_2O and CO_2:

$$CH_4 + 2O_2 \longrightarrow CO_2 + 2H_2O + heat$$
$$2C_6H_6 + 15O_2 \longrightarrow 12CO_2 + 6H_2O + heat$$

The amount of heat given off by the reaction is unique to each type of hydrocarbon. The normal measure of heat in the petroleum industry is the British thermal unit (Btu).

Definition. The amount of heat required to raise the temperature of one lb of water 1°F is equal to one Btu.

The table below gives the heating values for some commercial petroleum products:

Product	Higher Heating Value
Natural gas	1000-1050 BTU/scf
Ethane	66,000 BTU/gal
Propane	91,600 BTU/gal
Butane	103,300 BTU/gal
Distillate fuel	140,000 BTU/gal
No 6. Fuel (2.5% sulfur)	153,000 BTU/gal
No 6. Fuel (0.3% sulfur)	151,500 BTU/gal

*Table 22-1—*Heating Values

There are two types of heating values. The so-called higher heating value represents the gross amount of heat given off by burning it (the heat of combustion). The lower heating value takes into account what happens to the water created during the chemical reaction, burning. A portion of the heat of combustion is absorbed to vaporize the water. Not all that heat is recovered or usable as the heat goes through a furnace and out a stack. As a general rule, the heavier the fuel, the lower the ratio of hydrogen to carbon so the less H_2O formed during combustion. Typical results of this inability to recapture the heat of vaporization are measured by the thermal efficiency, the recoverable heat divided by the gross heating value. (Lower ÷ higher heating value).

Product	Thermal Efficiency, %
Natural gas	84
Propane	85
Distillate fuel	88
Coal	90

Table 22-2—Thermal Efficiency

Competitive Fuel Value Nomogram

A handy way to relate the values of the various products and their heating values is shown in the nomogram in Figure 22-1. That peephole diagram first appeared in the *Oil & Gas Journal* in 1972 and was updated to put higher scales on it in 1977 and then again in 1980.

As an example of nomogram use, assume you wish to find the fuel values equivalent to $2/MM Btu natural gas. First, go to the vertical scale on the right and find the horizontal line that intersects it at $2. Next, read along the horizontal line, right to left, the intersections on the other fuels scales: 13.2 cents per gallon (cpg) for ethane, 18.3 cpg for propane, 20.8 cpg for butane, etc.

Similarly, the value of any fuel can be equated to any other fuel by extending a straight line from the vertex on the left through the first fuel's value until it intersects the second fuel's scale.

For example, by laying a straight edge from the vertex through the high sulfur No. 6 fuel oil scale at $10/bbl, you can see the equivalent

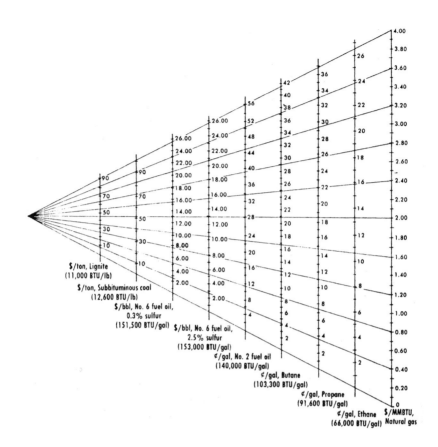

Fig. 22-1—Equivalent Fuel Values for Selected Hydrocarbons

values are $34/ton for lignite, 22 cpg for distillate fuel, 14 cpg for propane, and $1.57/MM Btu for natural gas.

Exercises

1. Which is the better buy for a refinery manager to use as refinery fuel, No. 2 fuel oil at $16/bbl, No. 6 fuel at $10/bbl, or natural gas at $2.10/MMBtu?

Answers and Solutions

What is the answer?...
In that case, what is the question?

-Alice B. Toklas,
What is Remembered,
Gertrude Stein

c h a p t e r   t w o

1. a. Calculate the cumulative % volumes and plot.
    b. The Oklahoma Sweet has 9.7%. The Heavy California crude only starts to boil somewhere between 260° and 315°, so it does not have the full range of naphtha. The volume is only 4.2%.

2. Because of the peculiar formula for API gravity, you can't add them and divide by two. You have to convert the API gravities to specific gravities, add them, divide by two, then convert them back to API gravity:

    $11° = 0.9930$ sp. gr.
    $50° = 0.7796$ sp. gr.
    $\qquad 1.7726 \div 2 = 0.8863 = 28.15°$

c h a p t e r   t h r e e

1. a. cooler
    b. batch, continuous
    c. bubble cap . . . downcomer
    d. bubble cap
    e. end point . . . initial boiling point
    f. decreases
    g. decreases . . . increases

2. There are several ways to solve the problem. One simplifying assumption you can make is that the distillation curve is a set of straight lines between the cut point/% volume plots. Then calculate a composite distillation curve by figuring the volume in each cut from the two crudes.

Volume-M B/D	
IBP-113	31.5
113-260	25.8
260-315	35.1
315-500	21.5
500-750	37.6
750-1000	20.0
1000+	28.5
	200.0

The next step can be done either by algebra or by plotting the distillation curve. The algebraic method is as follows. The IBP of the SRLGO is 525. That means that some of the 500-750° cut goes into jet fuel. How much? Let x be the amount. Then,

$$\frac{525 - 500}{750 - 500} = \frac{x}{37.6}$$

$$x = 3.76$$

That means the 525-750 cut has in it:

$$37.6 - 3.76 = 33.84 \ M \ B/D$$

To get only 20 MB/D, starting at 525(, let y = the end point. Then,

$$\frac{y - 525}{750 - 525} = \frac{20.0}{33.84}$$

$$y = 658°F$$

c h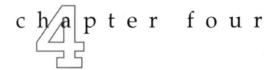a p t e r   f o u r

1. a.   distillation...cracking
    b.   lower
    c.   lower
    d.   light flashed distillate, heavy flashed distillate, and flasher bottoms
    e.   increase, lower, lower

2. a. To get 35 M B/D flasher bottoms, start at the "bottom of the barrel" on the distillation curve and work up. In problem 2, chapter 3, the volume of 1000°+ material is 28.5 M B/D. So, all that must be flasher bottoms. The 750°-1000° cut (20 M B/D) will cover the remainder of the requirement of 35.0-28.5 = 6.5 M B/D. Calculate as follows:

Let x be the IBP of the flasher bottoms. (Also knows as the flasher temperature and the end point of the flasher tops). Then,

$$\frac{1000 - x}{1000 - 750} = \frac{6.5}{20.0}$$

$$x = 918.75°$$

Of course the operations manager would probably be fired for doing this because flasher tops used as feed to the cat cracker is worth considerably more than asphalt, so he just downgraded the profitability of the refinery.

b. The flasher tops come from the 750-1000° cut, and go from 800-918.75.

Let y = Volume of flasher tops. Then,

$$\frac{y}{20} = \frac{918.75 - 800.0}{1000.0 - 750}$$

$$y = 9.5 \ M \ B/D$$

chapter five

1. Any way you build C_3H_8, the three carbons are connected to each other in the same fashion: two of them have three hydrogens and a carbon attached, one has two hydrogens and two carbons. The fact

that you can draw on paper one of the carbons attached at right angles to the others doesn't mean that in nature the molecule looks any different.

2. There's only one structure for iso-butane, a branch off the middle of three carbons.

```
          H   H   H
          |   |   |
      H - C - C - C - H
          |   |   |
          H   |   H
          H - C - H
              |
              H
          Isobutane
```

There are two isomers of pentane. The first iso-pentane has a branch off either of the two inside carbons in the string of four. It doesn't matter which one-they're symmetrical if you turn them around. The other iso-mer, neopentane, has four carbon branches off the inside carbon.

```
      H   H   H   H
      |   |   |   |
  H - C - C - C - C - H
      |   |   |   |
      H   H   |   H
          H - C - H
              |
              H
        Isopentane
```

```
                H
                |
            H - C - H
      H         |         H
      |         |         |
  H - C ------- C ------- C - H
      |         |         |
      H         |         H
            H - C - H
                |
                H
           Neopentane
```

Drawing the carbons slightly cock-eyed can best depict the one isomer of butylene. Then, no matter how you rotate it, it looks the same.

Isobutylene

```
       H    H
        \  /
         C
       /  \
      H    \      H
            \    /
             C = C
            /    \
      H    /      H
       \  C
        \/ \
        H   H
```

3. Paraffins, Olefins, Naphthenes, and Aromatics.

4. The three types of xylene, called para, meta, and orthoxylene, depend on where the methyl radicals are attached.

5. It doesn't matter where a single methyl radical is attached since the benzene ring is symmetrical.

c h a p t e r   s i x

1. A riser cracker is set up to do all the cracking in the riser that leads up to the disengagement vessel. A fluid cat cracker uses a catalyst that moves up the riser into the disengagement vessel or the reactor with the cat feed. Riser crackers are fluid cat crackers and both are cat crackers in general.

2. a. catalyst...hydrocarbon
    b. coke or carbon...air...CO...CO_2
    c. Heavy straight-run cuts...gasoline
    d. Distilling column...flasher
    e. Olefins
    f. Cycle oil...recycled to extinction

3. From the question on Flashing, flasher tops, with cut points 800-918.75°F, are 9.5 M B/D. The cut points of the straight-run heavy gas

oil must be the end point of the straight-run light gas oil, 658°F, (from the question on Distilling) and the initial boiling point of the straight-run residue, 800°F. The nasty job of calculating the volume of this 658°-800°F cut goes as follows:

Volume of 500-750°F = 37.6	(Distilling problem)
500-525°F = 3.76	(Distilling problem)
525-658°F = 20.0	(Distilling problem)

Therefore,

Volume of 658-750°F = 37.6 - 20 - 3.76
$$= 13.84$$

Volume of 750-1000°F = 20.0	(Distilling problem)
of 800-1000°F = 16.0	(Flasher problem)

Therefore,

Volume of 750-800°F = 4.0

Therefore,

Volume of straight-run heavy gas oil = 658-800°F = 13.84 + 4.0 = 17.84

Yield of cat light gas oil is 12%; cat cracker feed is the straight-run heavy gas oil and flasher tops, so the cat light gas oil volume is 0.12 x (17.84 + 9.5) = 3.28 MB/D.

4. (1) Decrease the initial boiling point or (2) increase the end point of the cat light gas oil; (3) increase the feed to the cat cracker by increasing the crude distilling column feed rate or (3) changing the cut points of the straight-run heavy gas oil and the flasher tops; (4) increasing the cycle oil volume by lowering the end point of the cat heavy gas oil; (5) change the operating conditions in the cat cracker reactor, or (6) the regenerator to change the cracking yields.

chapter seven

7

1. a. sats . . . cracked
    b. lean oil . . . fat oil
    c. sponge oil

2. Methane—Refinery fuel
    Ethane—Refinery fuel or chemical feed
    Propane—Commercial fuel or chemical feed
    Normal butane—Motor gasoline blending
    Iso-butane—Aklylation feed
    Propylene—Alkylation
    Butylenes—Alkylation
    Ethylene—Refinery fuel or chemical feed
    Hydrogen—Hydrotreating

3. One of the forms of butylenes and the iso-butylene boils at lower
    temperatures than normal butane; the other two normal butylenes
    boil at higher temperatures. The butylenes get split from each other
    to get the normal butane out.
    One distilling column configuration is in diagram 1. Several other
    schemes are possible as well. The important factor is understanding
    what goes overhead, what doesn't.

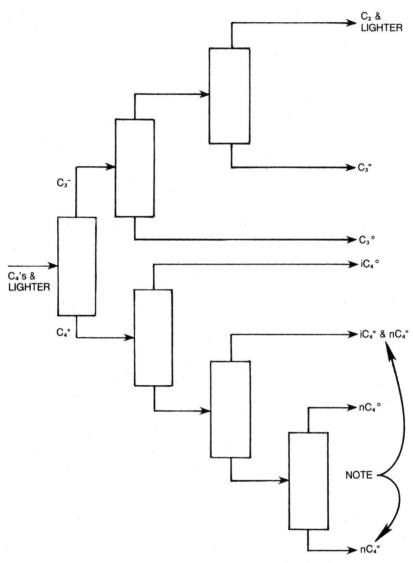

Diagram 1—Distillation Column Set-up to Segregate Cracked Gases

4. Diagram 2

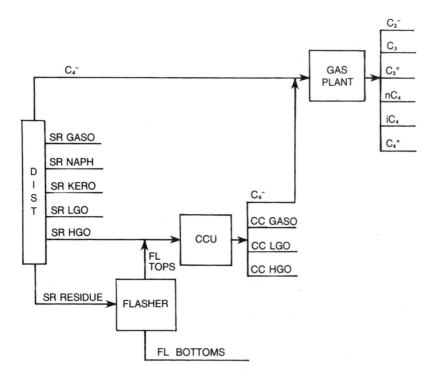

Diagram 2—Crude Distilling Column, Flasher, Cat Cracker and Gas Plant

c h a p t e r   e i g h t

8

1. a. cracking

   b. sulfuric acid or hydrofluoric acid

   c. cooler, reactor, acid separator, caustic wash, and fractionators

   d. iso-heptane and iso-octane

   e. octane number...vapor pressure...sulfur...benzene...olefins...
      gasoline blending component

2. First iC_4° needed for

   $C_3^=$ alkylation: 3000 x .25 x 1.6 = 1200 B/D

   $C_4^=$ alkylation: 3000 x .30 x 1.2 = 1080

                                  ———

                                  2280

less the 300 B/D in the feed is 1980 B/D

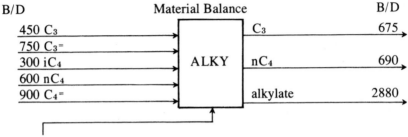

B/D	Material Balance		B/D
450 C3		C3	675
750 C3=			
300 iC4	ALKY	nC4	690
600 nC4			
900 C4=		alkylate	2880

1980 iC4 (from the Sats Gas Plant)

Then, alkylate production: yields

From $C_3^=$:   3000x0.15x1.8   = 1350 B/D

      $C_4^=$:   3000x0.30x1.7   = 1530

            Total alkylate   2880 B/D

The propane and butane out of the plant equals the propane and butane in plus what gets created during the process.

c h a p t e r   n i n e

9

1. a.   octane
    b.   platinum
    c.   aromatics
    d.   hydrogen
    e.   methane, ethane, propane, and butanes
    f.   hydrogen
    g.   yield...octane number...butanes and lighter

2. The values of the yield are computed as follows.

From the chart:	91 Oct.	95 Oct.	100 Oct.
Reformate Yield % Volume	94.0	91.0	81.5
C_4 and lighter % Vol.	8.7	11.0	24.8
Reformate Value, ¢/gal.	100.0	104.0	109.0
C_4 and lighter value, ¢/gal.	50.0	50.0	50.0
Reformate value	$9,400	$9,464	$8,835
C_4 and lighter value	$435	$550	$1,240
Total Value	$9,235	$10,014	$10,123.5

The total value keeps going up—only slightly, but up—as the octane number is raised despite the falling yield of reformate. So it makes sense to increase severity.

3. Paraffins to Isoparaffins:

Normal heptane   Isoheptane

Paraffins to Naphthenes:

Normal hexane   Cyclohexane

Naphthenes to Aromatics

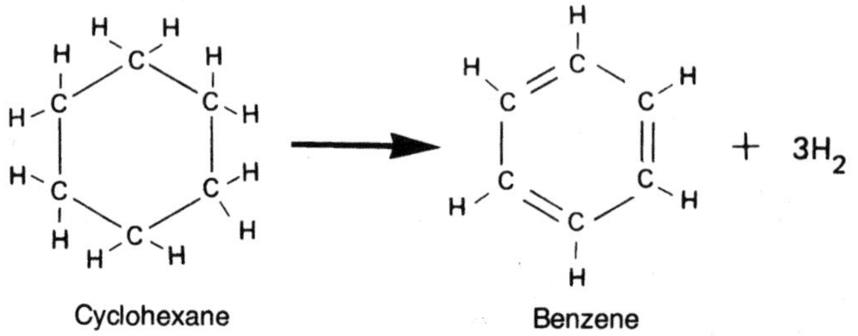

Cyclohexane   Benzene

$+ \ 3H_2$

Naphthenes Crack to Butanes and Lighter

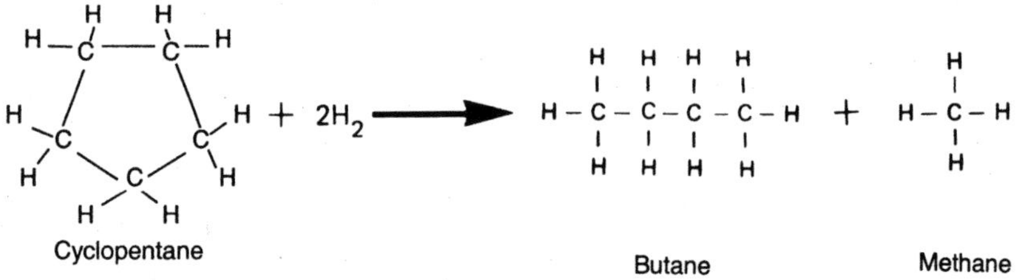

Cyclopentane

$+ \ 2H_2$

Butane   Methane

Side Chains Crack off Aromatics

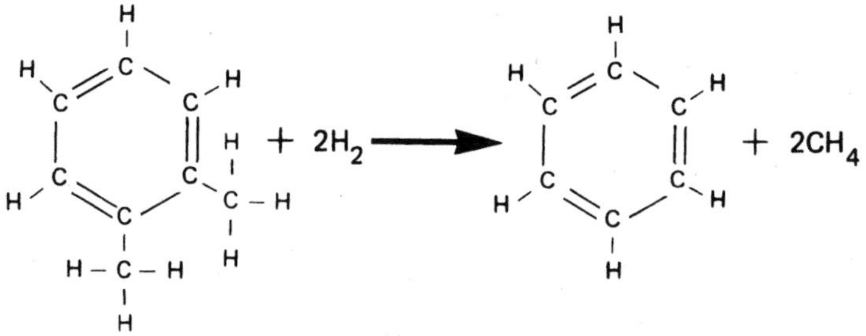

$+ \ 2H_2$ $+ \ 2CH_4$

4.   Platinum plus reformate gives platformate

5.   Diagram 3

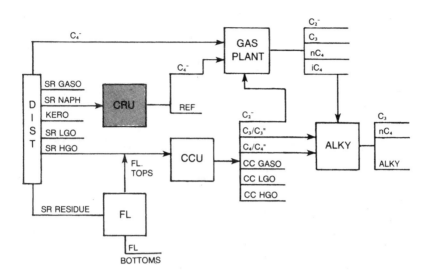

Diagram 3—**Refinery Operation with Catalytic Reformer**

c h a p t e r t e n

1. a. Catalyst
    b. Coke and Residue
    c. Cooled with a recycle steam
    d. "Cooked" until it cokes
    e. Low or poor
    f. Sponge coke...green coke...calcinable coke...
        anode grade coke...needle coke
    g. Cracked . . . olefins
    h. Furnace, distillation column, and reactor...drum

2. Flasher bottoms are 35 MB/D

    Coke yield $= \dfrac{35{,}000 \times 350 \times 0.30}{2000} = 1837.5$ tons/day

3. Diagram 4

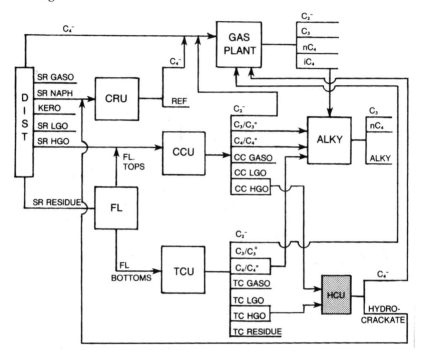

Diagram 4—Thermal Cracker Added to Refinery Process

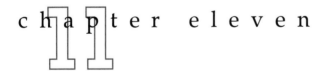

c h a p t e r   e l e v e n

1.

	Hydrocracking	**Cat Cracking**	**Thermal Cracking**
Feed	gas oils	heavy straight-run gas oil	CCHGO flasher bottoms
Process promoter	catalyst, hydrogen	catalyst	heat
Product quality	paraffinic, naphthenic	naphthenic, aromatic olefinic	paraffinic, naphthenic, aromatic, olefinic

2. Cat cracker gas oils make good feed to hydrocrackers.
   Hydrocrackate makes good feed to reformers.

3. Diagram 5

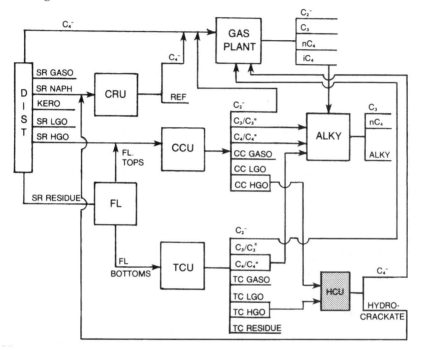

Diagram 5—Refinery Flow Diagram with Hydrocracker

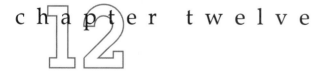

c h a p t e r   t w e l v e

1. *Vapor pressure* is a measure of the surface pressure necessary to keep a liquid from vaporizing.
   RVP stands for Reid Vapor Pressure and is the numerical result of measuring vapor pressure using Reid's method.

Power stroke is the downward motion of a piston that occurs after ignition as the fuel combusts and expands.

Vapor lock is the phenomenon of insufficient gasoline flow from a fuel pump due to its inability to pump the liquid-gas mixture that results from low pressure or high temperature.

Pressuring agent is the hydrocarbon, usually normal butane, used to bring gasoline blends up to an acceptable vapor pressure.

Knocking is the pre-ignition of gasoline in a cylinder during the compression stroke.

Driveability is achieved by smooth burning of gasoline during the power stroke and is enhanced by having a full range of hydrocarbons in the gasoline, which burn gradually over the whole stroke.

Oxygenates are ethers or alcohols used as a gasoline blending components. They have at least one oxygen atom connected to the other carbons and hydrogens.

Compression ratio is a measure of the amount of compression that takes place in an engine's cylinder and is equal to the volume of the cylinder at the bottom of the intake stroke to the volume at the top of the compression stroke.

RON and *MON* are measures of octane numbers under conditions simulating mild and severe conditions, respectively.

Leaded gasoline is gasoline that has TEL added to boost the octane number.

Octane enhancement is the reaction to the addition of high octane components such as oxygenates and measured by MON or RON.

2. a. To get the amount of butane necessary to meet the RVP spec:

	Barrels	RVP	MON	RON
Alkylate	3000	4.6	95.9	97.3
Reformate	2500	2.8	84.4	94.0
Hydrocrackate	6,000	2.5	73.7	75.5
CC Gasoline	3,600	4.4	76.8	92.3
Subtotal	15,100	3.4		
Normal butane	x	52.0	92.0	93.0

$$(15,100)(3.4) + 52.0x = (15,100 + x)(10.0)$$
$$x = 2,373 \text{ bbls to meet the RVP spec}$$

Blend, including the butane	17,473	10.0	82.2	87.7

b. To get the amount of MTBE to meet the octane specs:

	Barrels	RVP	MON	RON
Pressured gasoline	17,473	10.0	82.2	87.7
MTBE	y	8.0	103.0	115.0

To blend up to the MON spec, let y_1 = the amount of MTBE to meet the MON spec

$$(1473)(82.2) + 103.0\, y_1 = (17,473 + Y1)(84.0)$$
$$y_1 = 1,655 \text{ bbls}$$

To blend up to the RON spec, let Y_2 = the amount of MTBE to meet the RON spec

$$(17,473)(87.7) + 115.0\, y_2 = (17.473 + y2)(90.0)$$
$$y_2 = 1,608 \text{ bbls}$$

So 1,655 bbls of MTBE have to be added to meet the MON spec and the extra RON is an octane giveaway.

c. When the MTBE is added in, the vapor pressure of the gasoline blend goes down a tad, making room for more butane. To get the balance between butane and MTBE, you have to use a two equation/two unknown calculation.

d. MTBE boils at a single temperature. Too much MTBE in the blend will give a flat spot on the gasoline distillation curve. The gasoline may not have sufficient distribution of compounds to give an even burn over the whole power thrust of the piston. The result might be a rough burning gasoline.

c h a p t e r   t h i r t e e n

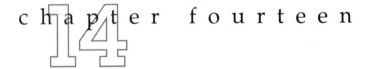

1. Gasoline ignites when the spark plug gives off a spark. Diesel ignites when the fuel is injected into the cylinder and hits the heated, com pressed air.

    Octane numbers and cetane numbers are measured in more or less the same way, but use different engines and different test fluids (iso-octane and normal heptane for gasoline and cetane and alpha-methylnaphthalene for diesel).

    Aromatics favor gasoline octane number but have low cetane numbers. Iso-paraffins and naphthenes favor octane numbers. Normal paraffins have low octane numbers and high cetane numbers.

2. Diesel has a cetane specification on it and in many places a max sulfur spec. Furnace oil has no cetane spec and in many places no sulfur spec.

c h a p t e r   f o u r t e e n

1. Asphaltenes.

2. Not all crudes are asphalt-suitable.

3. A ship can have a boiler with a steam-driven turbine or no boiler and a marine diesel engine.

c h a p t e r   f i f t e e n

1.

	Hydrotreating	DEA Removal	Claus Plant	SMR
H_2S	P	F, P	F, I	—
S	—	—	P	—
CO	—	—	—	I
CO_2	—	—	—	I
CH_4	—	—	—	F, P
H_2	F	—	—	F, P
O_2	—	—	F	—
SO_2	—	—	I, P	—

P-Product
F-Feed
I-Internal Steam

c h a p t e r   s i x t e e n

1. Refiners include a BI plant in their configuration because they are short of iso-butane needed for their alky plant. They have a C_5/C_6 isom plant because they need to upgrade the octane numbers of their C_5's and C_6's, most of which are in their straight run gasoline.

2. To get the normal hexane split out for recycling to extinction, you must add a deisohexanizer downstream of the C_5/C_6 splitter and route the bottoms to the reactor. You'd probably want to then configure all three fractionators so all the iso-hexane is split out of the feed to avoid passing it through the reactor.

3. The refiner can use an isomerization plant to make iso-butane then a dehydrogenation plant to make iso-butylene; reverse the order and use a dehydrogenation plant to make normal butylene and then a isomerization plant to make iso-butylene; or just sell the normal butane and buy some iso-butylene.

c h a p t e r   s e v e n t e e n

1. | | |
   |---|---|
   | MTBE | Methanol and iso-butylene |
   | ETBE | Ethanol and iso-butylene |
   | TAME | Methanol and iso-pentene |
   | TAEE | Ethanol and iso-pentene |
   | THxME | Methanol and iso-hexene |
   | THxEE | Ethanol and iso-hexene |
   | THpME | Methanol and iso-heptene |
   | THpEE | Ethanol and iso-heptene |

2. The signature for ether is embedded oxygen with some organic radicals, R_1 and R_2, which could be the same radical, on either side. The chemical notation is $\mathbf{R_1\text{-}O\text{-}R_2}$.

c h a p t e r   e i g h t e e n

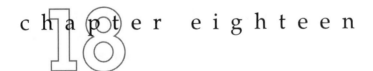

1. The solvent; benzene raffinate, is what's left over after the benzene has been taken out.

2. It selectively dissolves the target compound, readily separates itself from the original mixture, and can easily be separated from the tar get compound by distillation.

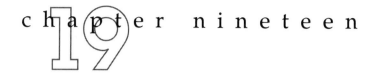

c h a p t e r   n i n e t e e n

1. Ethane: $1 \times 10° \div (0.77 \times 3.2 \times 42 \times 365)$ = 26 MB/D
    Propane: $1 \times 10° \div (0.40 \times 4.24 \times 42 \times 365)$ = 38 MB/D
    Naphtha: $1 \times 10° \div (0.23 \times 6.4 \times 42 \times 365)$ = 44 MB/D
    Gas Oil: $1 \times 10° \div (0.18 \times 7.3 \times 42 \times 365)$ = 50 MB/D

2. Let x equal the feed in b/d of the 70/30 mix. To make 500 MM lb of
    ethylene, the feed rate is as follows:

    Calculate the weighted average ethylene yield from the ethane and
    propane. The yield from ethane is x times the percent ethane times the
    pounds per gallon ethane times the yield of ethylene from ethane times
    42 gal per barrel times 365 days per year. Same for ethylene from propane.

    From ethane   From propane   Ethylene
    $(X)(0.7)(3.2)(.77)(42)(365)$ + $(X)(0.30)(4.24)(0.40)(42)(365)$ = 500×10^6
                    x = 14.6 mb/d of ethane/propane feed

    The propylene production was:

    From ethane   From propane   Propylene
    $(14.6)(0.7)(3.2)(42)(365)(0.1)$ + $(14.6)(0.3)(4.24)(42)(365)(0.18)$ = 56.5×10^6

    To find out how much ethane and propane must be cracked to
    make 500 MM lb of ethylene but only 20 MM lb of propylene,
    use simultaneous equations:

    Let y equal the ethane feed rate in b/d
        z equal the propane feed rate in b/d

Then:

	From ethane	From propane
Ethylene:	$500 \times 10^6 = (y)(3.2)(42)(365)(0.77) + (z)(4.24)(42)(365)(0.40)$	
Propylene:	$20 \times 10^6 = (y)(3.2)(42)(365)(0.01) + (z)(4.24)(42)(365)(0.18)$	

$$500 \times 10^6 = 37{,}773\,y + 26{,}000\,z$$
$$20 \times 10^6 = 490\,y + 11{,}700\,Z$$

$y \qquad = 12.4$ MB/D ethane

$z \qquad = 1.2$ MB/D propane

13.6 MB/D of 91/9 mix

c h a p t e r   t w e n t y

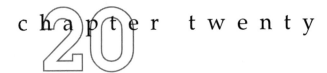

1. If the market is awash with light oils but the residual fuel market is very strong, the profitability of a simple refinery rises and the very complex refinery falls. That could happen if a natural gas shortage coincided with a recession.

2. Since a coker converts heavy oil to light oil, the difference in price between the two is what pays for the capital invested in a coker.

3. Simple refinery margins will rise, complex refinery margins will fall, and very complex refining margins will fall even faster.

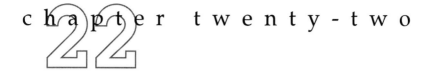

chapter twenty-two

1. No. 2 fuel oil at $16/bbl is a much better buy than the low sulfur No. 6 fuel oil or propane. Of course you have to assume the plant has the facilities to burn all three. Generally the facilities for No. 6 fuel oil and for propane are more expensive to build and for No. 6 fuel oil more expensive to operate (since it needs to be heated to move and burn it).

Glossary

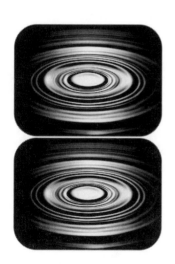

The world's no blot for us.
Nor blank; it means intensely, and means good:
To find its meaning is my meat and drink.

-Fra Lippo Lippi,
Robert Browning

Absorbent. The material that can selectively remove a target constituent from another compound by dissolving it.

Absorption. A variation on fractionation. In a distilling column the stream to be separated is introduced in vapor form near the bottom. An absorbing liquid, called lean oil, is introduced at the top. The lean oil properties are such that as the two pass each other, the lean oil will selectively absorb components of the stream to be separated and exit the bottom of the fractionator as fat oil. The fat oil is then more easily separated into the extract and lean oil in conventional fractionation.

Adsorbents. Special materials like activated charcoal, alumina, or silica gel, used in an adsorption process, that selectively cause some compounds, but not others, to attach themselves mechanically as liquids.

Adsorption. A process for removing target constituents from a stream by having them condense on an adsorbent, which is then taken off line so the target constituents can be recovered.

Alkylation. 1. A refining process in which propylene or butylene are reacted with iso-butylene to form a high octane gasoline blending component, alkylate; 2. Any chemical process in which an alkyl radical (a radical derived from a paraffin) is connected to an organic compound.

API gravity. An arbitrary scale used for characterizing the gravity of a petroleum product. The degrees API (written °API) are related to specific gravity scale by the formula: $°API = [141.5/(sp.gr. @ 15°C/15°C)] - 131.5$

API. The American Petroleum Institute, an association which, among many other things, sets technical standards for measuring, testing, and other types of handling petroleum.

Aromatics. 1. A group of hydrocarbons characterized by having at least one benzene ring type structure of six carbon atoms somewhere in the molecule. The simplest is benzene itself, plus toluene and the xylenes. Aromatics in gas oils and residuals can have many, even scores, of rings. 2. The three aromatics compounds, benzene, toluene, and xylene.

Asphalt. 1. A heavy, semi-solid petroleum product that gradually softens when heated and is used for surface cementing. Typically brown or black in color, it is composed of high carbon-to-hydrogen hydrocarbons plus some oxygen. It occurs naturally in crude oil or can be distilled or extracted. 2. The end product used for area surfacing consisting of refinery asphalt mixed with aggregate.

Asphaltenes. Polyaromatic materials in heavy residues characterized by not being soluble in aromatic-free low-boiling point solvents. They are soluble in carbon disulfide.

Associated natural gas. Natural gas that is dissolved in crude in the reservoir and is co-produced with the crude oil.

Barrel. A standard of measurement in the oil industry; equivalent to 42 US gallons, 35 Imperial gallons, or 159 liters.

Benzene (C_6H_6). A chemical consisting of a six-carbon ring connected by with double and single bonds. Benzene has excellent octane characteristics but is carcinogenic and therefore its content in gasoline is limited severely by regulation. Benzene is used in a large number of chemical processes including styrene and detergents.

Bitumen. The British term for asphalt.

Black oil. Residual fuel or more generally the very heavy residues in the refinery.

Blown asphalt. A special grade of asphalt made by oxidizing flasher bottoms by blowing heated air through it.

Boiling point. The temperature at which a liquid will boil. (See end point and initial boiling point.)

Boiling range. The lowest temperatures between which a hydrocarbon will begin to boil and completely vaporize.

Bottoms. 1. The product coming from the bottom of a fractionation column. 2. The liquid level left in a tank after it has been pumped "empty" and the pump loses suction.

Bright stock. A lubricating oil component with high viscosity. Usually made by vacuum distillation of paraffinic crude oil, deasphalting, and solvent extraction of aromatics.

British thermal unit (BTU). A standard measure of energy: the quantity of heat required to raise the temperature of 1 pound of water by 1°F.

BS&W. The bottom sediment and water that settle out of petroleum stored in a tank.

Bubble cap tray. The trays in a fractionator consisting of a plate with holes and bubble caps. The latter cause the vapor coming from the bottom to come in intimate contact with the liquid sitting on the tray.

Bunker fuel. Fuel oil or diesel used as ship fuel. Originally coal stored in the ship's bunkers.

Butadiene (CH_2=CHCH=CH_2). A colorless gas resulting from cracking processes. Traces result from cat cracking; made on-purpose by catalytic dehydrogenation of butane or butylene and in ethylene plants using butane, naphtha, or gas oil as feeds. Butadiene is principally used to make polymers like synthetic rubber and ABS plastics.

Butane (C_4H_{10}). Commercial butane is typically a mixture of normal butane and isobutane, predominantly normal. To keep them liquid and economically stored, butane must be maintained under pressure or low temperatures.

Butylene (C_4H_8). Hydrocarbons with of several different isomers in the olefin series, used as raw materials for making gasoline blending components in an alkylation plant or solvents and other chemicals.

Catalyst. A substance present in a chemical reaction that will promote, or even cause, the reaction, but not take part in it by chemically changing itself. Sometimes used to lower the temperature or pressure at which the reaction takes place or speed it up.

Catalytic cracking. A central process in refining in which heavy gas oil range feeds are subjected to heat in the presence of a catalyst and large molecules crack into smaller molecules in the gasoline and surrounding ranges.

Catalytic reforming. The process in refining in which naphthas are changed chemically to increase their octane numbers. Paraffins are converted to iso-paraffins and naphthenes; naphthenes are converted to aromatics. The catalyst is platinum and sometimes palladium.

Caustic soda. The name used for sodium hydroxide (NaOH); used in refineries to treat acidic hydrocarbon streams to neutralize them. The term is derived from the corrosive effect on skin.

Cavern. Underground storage either leached out of a natural salt dome or mined out of a rock formation.

Cetane number. A measure of the ignition quality of diesel fuel and set by the percent of cetane in a mixture with alpha-methylnaphthalene.

Centipoise. A measure of viscosity, related to centistokes by adjusting for density.

Centistoke. A measure of viscosity.

Coke. 1. A product of the coking process in the form of solid, densely packed carbon atoms. Various forms of coke include green coke, a run of the mill coke from most cokers; sponge coke, the same as green coke and notable by its fine, sponge-like structure; calcinable coke, a high grade of coke that is suitable for making industrial products; needle coke, a very high grade of coke characterized by crystalline structure. 2. Deposits of carbon that settle on catalysts in cat crackers, cat reformers, hydrocrackers, and hydrotreaters and degrade their effectiveness.

Coker. A refinery process in which heavy feed such as flasher bottoms, cycle oil from a cat cracker, or thermal cracked gas oil is cooked at high temperatures. Cracking creates light oils; coke forms in the reactors and needs to be removed after they fill up.

Compression ratio. The ratio of volumes in a cylinder when the piston is at the bottom of the stroke and the top of the stroke, giving a measure of how much the air or air/fuel mixture is compressed in the compression stroke.

Condensate. 1. The relatively small amount of liquid hydrocarbon, typically C_4's through naphtha or gas oil, that gets produced in the oil patch with unassociated gas, 2. The liquid formed when a vapor cools.

Cracked gas. The C_4- stream coming from a cat cracker, coker, or thermal cracker, containing olefins in addition to the saturated paraffins.

Cracked gas plant. The set of columns and treaters in a refinery that handle separation and treating of the cracked, olefinic gases.

Cracking. Refining processes in which large molecules are broken into smaller molecules of uneven sizes. It can be promoted by heat and pressure alone, such as in thermal cracking or coking, or enhanced by the use of catalysts as in cat cracking or hydrocracking.

Cut. A refinery term referring to a stream obtained from a fractionation unit with a specific initial boiling point and end point.

Cutter stock. Diluent added to residue to meet residual fuel specifications for viscosity and perhaps sulfur content. Typically cracked gas oil.

Cyclic compounds. (See ring compounds).

Deasphalting. A process in which the asphaltic constituents of a heavy residual oil are separated by mixing with liquid propane. Everything will dissolve in the propane but the asphaltics, which can then be easily removed.

De-coking. The process of removing coke from catalysts in a cat cracker, cat reformer, hydrocracker, or hydrotreater. Usually heated air will oxidize the coke to carbon monoxide or carbon dioxide.

Delayed Coker. A process unit in which residue is cooked until it cracks to coke and light products.

Diene. Same as a di-olefin.

Diesel. 1. An internal combustion engine in which ignition occurs by injecting fuel in a cylinder where air has been compressed and is at a very high temperature, causing self-ignition. 2. Distillate fuel used in a diesel engine.

Di-olefin (C_nH_{2n-2}). A paraffin-type molecule except that it is missing hydrogen atoms that causes it to have two double bonds somewhere along its chain.

Distillate. 1. The liquid obtained by condensing the vapor given off by a boiling liquid; 2. Any stream, except the bottoms, coming from a fractionator; 3. The products or streams in the light gas oil range such as straight run light gas oil, cat cracked light gas oil, heating oil, or diesel.

Distillation range. (See boiling range.)

Distillation. A separation process which results in products with different boiling ranges. Distillation is carried out in a way that the materials being separated are not subjected to conditions that would cause them to crack or otherwise decompose.

Distributor. A device in a vessel that disperses either liquid or vapor to promote better circulation.

Dry gas. Natural gas or refinery gas streams that are in the C_4 and lighter range.

Effective cut points. Cut points that can be considered a clean cut, ignoring any tail ends.

End point. The lowest temperature at which virtually 100% of a petroleum product will boil off to vapor form.

Ethane (C_2H_6). A colorless gas, a minor constituent of natural gas and a component in refinery gas that, along with methane, is typically used as refinery fuel. An important feedstock for making ethylene.

Ethylene (C_2H_4). A colorless gas created by cracking processes. In refineries it is typically burned with the methane and ethane. In chemical plants it is made in ethylene plants and is a basic building block for a wide range of products including polyethylene and ethyl alcohol.

Extract. The target constituent in a solvent extraction process. (See solvent extraction).

Fixed bed. A place in a vessel for catalyst through or by which feed can be passed for reaction; as opposed to a fluid bed, where the catalyst moves with the feed.

Flash chamber. A wide vessel in a vacuum flasher, thermal cracking plant, or similar operation into which a hot stream under pressure is introduced, causing the lighter fractions of that stream to vaporize and leave by the top.

Flash point. The lowest temperature at which any combustible liquid will give off sufficient vapor to form an inflammable mixture with air. Flash points are used to specify the volatility of fuel oils, mostly for safety reasons.

Flue gas. Gas from the various furnaces going up the flue (stack).

Fluid bed. See fixed bed

Fluid cat cracking. The most popular design of cat cracking in which a powdery catalyst that flows like a fluid is mixed with the feed and the reaction takes place as the feed/catalyst is in motion.

Fractionation. The general name given to the process for separating mixtures of hydrocarbons or other chemicals into separate streams or cuts or fractions.

Freeze point. The temperature at which crystals first appear as a liquid is cooled, which is especially important in aviation fuels, diesel and furnace oil.

Fresh feed rate. The feed going into a reaction without counting the amount of recycling of any reaction products.

Fuel oil. Usually residual fuel but sometimes distillate fuel.

Furnace oil. A distillate fuel made of cracked and straight run light gas oils, primarily for domestic heating because of its ease of handling and storing.

Gain. The volumetric expansion resulting from the creation of lighter, less dense molecules from heavier, compact molecules, even given the same weight, before and after.

Gas cap. An accumulation of natural gas in at the top of a crude oil reservoir. The gas cap often provides the pressure to rapidly evacuate the crude oil from the reservoir.

Gas oil. A fraction of hydrocarbon in the range of approximately 450–800°F. Sources are crude oil distilling and cracking.

Gasoline. A light petroleum product in the range of approximately 80–400°F for use in spark-ignited internal combustion engines.

Gum. A complex, sticky substance that forms by the oxidation of gasolines, especially those stored over a long period of time. Gum fouls car engines, especially the fuel injection ports.

Heat exchanger. An apparatus for transferring heat from one liquid or vapor stream to another. A typical heat exchanger will have a cylindrical vessel through which one stream can flow and a set of pipes or tubes in series in the cylinder through which the other can flow. Heat transfers through the tubes by conduction.

Heating oil. Any distillate or residual fuel.

HF alkylation. Alkylation using hydrofluoric acid as a catalyst.

Hydrocarbon. Any organic compound comprised of hydrogen and carbon, including crude oil, natural gas and coal.

Hydrocrackate. The gasoline range product from a hydrocracker.

Hydrocracking. A process in which light or heavy gas oils or residue hydrocarbons are mixed with hydrogen under high pressure and temperature and in the presence of a catalyst to produce light oils.

Hydrodesulfurization. A process in which sulfur is removed from the molecules in any refinery stream by reacting it with hydrogen in the presence of a catalyst.

Hydrogenation. Filling in with hydrogen the "free" places around the double bonds in an unsaturated hydrocarbon molecule.

Hydrophilic. An adjective meaning having an affinity for water. For example, bourbon has an affinity for water because it mixes easily with it. Olive oil is not hydrophilic because it does not. (It is hydrophobic.) A single molecule can have both a hydrophilic and hydrophobic sites, such as soap.

Hydrophobic. (See hydrophilic).

Hydrotreating. A process in which a hydrocarbon is subjected to heat and pressure in the presence of a catalyst to remove sulfur and other contaminants such as nitrogen and metals and in which some hydrogenation can take place.

Initial boiling point (IBP). The lowest temperature at which a petroleum product will begin to boil.

Isomerization. A refinery process in which compounds are changed to their isomer form using a catalyst. For example normal hexane to iso-hexane or normal butane to iso-butane.

Isomers. Two compounds composed of the identical atoms, but with different configurations, giving different physical properties.

Iso-octane (C_8H_{18}). The liquid used with normal heptane to measure the octane number of gasolines.

Jug. A salt dome storage cavern for hydrocarbon or chemicals.

Kerosene. A petroleum product made from crude oil with a boiling range of approximately 315-450°F, used as domestic heating oil.

Lean oil. (See Absorption process).

Light ends. In a refinery, the CH_4 and lighter gases.

Light oils. Generally gasoline, kerosene, and distillate fuels.

Liquefied petroleum gas (LPG). Propane and butane meeting market specifications.

Long residue. Straight run residue from a distilling unit.

Mercaptans. A group of sulfur-containing compounds found in some crude oils having a skunk-like odor. Mercaptans are manufactured and injected into natural gas and LPG as an odorant for safety purposes.

Methane (CH_4). A light, odorless, flammable gas that is the principal component of natural gas.

Methanol (CH₃OH). Methyl alcohol, also known as wood alcohol. Methanol can be made by the destructive distillation of wood or through a process starting with methane or a heavier hydrocarbon, decomposing it to synthesis gas, and recombining it to methanol.

Motor octane number (MON). One of two standard measures of gasoline knock, this one simulating more severe operating conditions.

Naphtha. Hydrocarbon fractions in the range of approximately 220-315°F. Naphtha is generally not suitable for direct blending to gasoline and is usually further processed in a cat reformer or used as feed to an ethylene plant.

Naphthenes. Hydrocarbons with saturated ring structures with a general formula C_nH_{2n}.

Naphthenic acids. Organic acids occurring in petroleum that contain a naphthene ring and one or more carboxylic acid groups. Naphthenic acids are used in the manufacture of paint driers and industrial soaps.

Natural gas. Naturally occurring gas consisting predominantly of methane, sometimes in conjunction with crude (associated gas), sometimes alone (unassociated gas).

Natural gasoline. A gasoline range product separated at a location near the point of production from natural gas streams and used as a gasoline blending component.

Non-associated gas. Natural gas that exists in a reservoir alone and is produced without any crude oil.

Octane number. An index measured by finding a blend of iso-octane and normal heptane that knocks under the identical conditions as the gasoline being evaluated. It is a measure of the resistance to ignition of the fuel without the aid of a spark plug. The higher the octane number, the more resistance to pre- or self-ignition. (See MON and RON).

Olefins. A class of hydrocarbons similar to paraffins, but that has two hydrogen atoms missing and a double bond replacing it. The general formula is C_nH_{2n} for mono-olefins and C_nH_{2n-2} for di-olefins, those having two sets of double bonds.

Organic compounds. Any compound that includes carbon (except carbon dioxide and some carbonates). Generally organic compounds can be classified as either aliphatics (straight chain compounds), cyclics (compounds with ring structures), and combinations of aliphatics and cyclics.

Paraffins. Straight chain hydrocarbons with the general formula C_nH_{2n+2}.

Penetration. A measure of the hardness and consistency of asphalt in terms of the depth that a special pointed device will penetrate the product in a set time and temperature.

Petroleum Coke. (See Coke).

Pitch. Residue coming from the bottom of a flasher.

Platformer. Archaic name for a reformer.

Pour point. The temperature at which an oil starts to solidify and no longer flows freely.

Precursor. Compounds that are suitable or susceptible to specific conversion to another compound. For example, methyl cyclopentane is a good precursor for making toluene in a cat reformer.

Propane (C_3H_8). A hydrocarbon gas that is a principal constituent of the heating fuel, LPG. Propane is used extensively for domestic heating and as a feed to ethylene plants.

Propylene (C_3H_6). A hydrocarbon in the olefin series resulting from cracking processes and used as alky plant feed or chemical feedstock.

Pyrolysis gasoline (Pygas). The gasoline created in an ethylene plant using gas oil or naphtha feedstocks. Sometimes called pygas, it has a high content of aromatics and olefins and some di-olefins.

Pyrolysis. Heating a feedstock to high temperature to promote cracking, as in an ethylene plant.

Quench. Hitting a very hot stream coming out of a reactor, with a cooler stream to stop immediately the reaction underway.

Radical. A group of atoms from a compound that moves unchanged to another compound. A methyl radical, written as $-CH_3$ can come from methane and attach it self to a benzene radical, $-C_6H_5$ forming toluene, $C_6H_5CH_3$.

Raffinate. The leftover from a solvent extraction process. See solvent extraction.

Reactor. The vessel in which the chemical reaction takes place.

Reboiler. A heat exchanger used towards the bottom of a fractionator to re-heat or even vaporize a liquid and introduce it several trays higher to get more heat into the column to improve separation.

Reflux. A heat exchanger, which takes vapor from the upper parts of a fractionator, cools it to liquefy it, and reintroduces it lower on the column. The purpose is to assure sufficient downward liquid flow meeting the rising vapor to improve separation.

Reformate. The high octane, primary product of reforming naphtha.

Reforming. See cat reforming or steam-methane reformer.

Regenerator. The vessel in a catalytic process where a spent catalyst is brought back up to strength before being recycled back to the process. An example is the cat cracker regenerator where coke is burned off the catalyst.

Reid vapor pressure (RVP). The pressure necessary to keep a liquid from continually vaporizing, as measured in an apparatus designed by Reid himself. Used as a standard measure for gasoline specifications.

Research octane number (RON). One of two standard measures of gasoline knock, this one simulating less severe operating conditions like cruising.

Residence time. The amount of time a hydrocarbon spends in a vessel where a reaction is taking place.

Residual fuel (Resid). Heavy fuel oil made from long, short, or cracked residue plus whatever cutter stock is necessary to meet market specifications.

Residue. The bottoms from a crude oil distilling unit, vacuum flasher, thermal cracker, or visbreaker. See long residue and short residue.

Ring compounds. Hydrocarbon molecules in which the carbon atoms form at least one closed ring such as naphthenes or aromatics. Also called cyclics.

Salt dome. A naturally occurring column of salt lying several hundred to thousands of feet below the surface. Many salt domes are suitable for leaching out and using as hydrocarbon storage.

Sats gas plant. The set of columns and treaters in a refinery that handle separation and treatment of the saturated gases.

Short residue. Flasher bottoms or residue from the vacuum flasher.

Sieve trays. A variation of the trays used in fractionating columns, consisting of perforated plates to allow vapor passage.

Smoke point. The maximum height of flame in millimeters at which kerosene will burn without creating smoke. Measured in a standard wicked lamp.

Solvent extraction. A fractionation process based on selective solubility. A liquid solvent is introduced at the top of a column. As it passes the feed, which enters near the bottom as a vapor, it selectively dissolves a target constituent. The solvent is then removed via the bottom of the column and put through an easy solvent/extract fractionation. From the top of the column comes a raffinate stream, the feed with the extract stripped out of the extract. Butadienes and aromatics are some products recovered by solvent extraction.

Sour crude. Crude typically containing 1.5% (by weight) or more sulfur.

Spent catalyst. Catalyst that has been through a reaction and is no longer as active because of deposits or other contaminants laid down on (in the case of solid) or mixed with (in the case of liquid) it.

Sponge oil. The liquid used in an absorption plant to soak up the constituent to be extracted. (See absorption or solvent extraction.)

Stability. A measure of the resistance of petroleum products to forming gums while in storage.

Stabilizer. A fractionator used to remove most of the light ends from straight run gasoline or natural gasoline to make them less volatile.

Steam cracking. The same as cat cracking, but specifically referring to the steam injected with the catalyst and feed to give the mixture lift up the riser.

Steam-methane reformer. A primary source of hydrogen in a refinery, this operating unit converts methane and steam to hydrogen, with by-product carbon monoxide and carbon dioxide.

Straight run. A product of distillation but no chemical conversion.

Sulfolane $(CH_2)_4SO_2$. A chemical used as a solvent in extraction and extractive distillation processes.

Surface area. The total area that a solid catalyst exposes to the feeds in a reaction. Surface area is enhanced in some catalysts like zeolytes by extensive microscopic pores.

Sweet crude. Crude typically containing 0.5% (by weight) or less sulfur.

Sweetening. The conversion of mercaptans in gasoline into non-smelly disulphides

Synthesis gas. The product of a reforming operation in which a hydrocarbon, usually methane, and water are chemically rearranged to produce carbon monoxide, carbon dioxide, and hydrogen. The composition of the product stream can be varied to fit the needs for hydrogen and carbon monoxide ratios at refineries or chemical plants.

Tail ends. Small amounts of hydrocarbon in a cut that vaporize slightly outside the effective initial boiling point and the effective end point.

Tar. Complex, large molecules of predominantly carbon with some hydrogen and miscellaneous other elements that generally deteriorate the quality of processes and apparatus.

Tetraethyl lead (TEL). A nearly extinct additive used to enhance the octane number of gasolines. TEL is outlawed in most countries as a health hazard.

Thermal cracking. A refinery process of cracking heavy streams such as flasher bottoms or cat cracked cycle oil into light products using high temperatures

Toluene ($C_6H_5CH_3$). One of the aromatic compounds used as a chemical feedstock, most notoriously for the manufacture of TNT, trinitrotoluene

Topped crude. Crude that has been run through a distilling unit to remove the gas oil and lighter streams. The so-called simple refineries that do this sell the long residue as residual fuel.

Unsaturated. A class of hydrocarbons similar to paraffins and naphthenes, but that has double bonds or triple bonds replacing the missing hydrogens.

Vacuum distillation. Distillation under reduced pressure in order to keep the temperature low and prevent cracking. Most often used to distill lubricant feedstocks.

Valve trays. Fractionator trays that have perforations covered by discs which operate as valves and allow the passage of varying amounts of vapor to flow upward.

Vapor lock. Atmospheric conditions of high temperature or low pressure can cause gasoline to vaporize in the fuel line, disabling the fuel pump and shutting down the engine due to vapor lock.

Vapor pressure. (See Reid vapor pressure).

Visbreaking. Mild thermal cracking aimed at producing sufficient middle distillates to reduce the viscosity of the heavy feed.

Viscosity. The measure of a liquid's resistance to flow.

Volatile. A hydrocarbon is volatile if it has a sufficient amount of butanes and lighter material to noticeably give off vapors at atmospheric conditions.

Volatiles. Oil patch nomenclature for butane and propane, and sometimes ethane.

Wet gas. Natural gas that has not had the C_4 and natural gasoline removed. Also the equivalent refinery gas stream.

White oil. Same as light oil. See black oil, the opposite.

Xylene ($C_6H_4(CH_3)_2$. One of the aromatics compounds. Xylene has a benzene ring and two methyl radicals attached and has three isomers, ortho-, para-, and meta-xylene. Used as a gasoline blending component or a chemical feedstock for making phthalic acids and resins.

Yield. Either the amount of a desired product or all the products resulting from a process involving chemical changes to the feed.

Zeolytes. Compounds used extensively as catalysts, made of silica or aluminum, as well as sodium or calcium and other compounds. Zeolytes come in a variety of forms - porous and sand-like or gelatinous and provide the platform for numerous catalysts. The solid zeolytes have extensive pores that give them very large surface areas. The precise control during fabrication of the pore sizes enables selected access to different size molecules during reactions.

Index

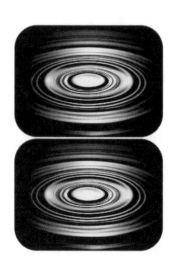

A

B

E

F

G

H

I

J

K

L

M

N

O

Q

R

T

U

V

WXYZ